D1712879

Sod Busting

How Things Worked

Robin Einhorn and Richard R. John, Series Editors

ALSO IN THE SERIES:

Sean Patrick Adams, *Home Fires: How Americans Kept Warm in the Nineteenth Century*

Ronald H. Bayor, *Encountering Ellis Island: How European Immigrants Entered America*

Bob Luke and John David Smith, *Soldiering for Freedom: How the Union Army Recruited, Trained, and Deployed the U.S. Colored Troops*

Sod Busting

How Families Made Farms on the Nineteenth-Century Plains

DAVID B. DANBOM

LINDENHURST MEMORIAL LIBRARY
LINDENHURST NEW YORK 11757

Johns Hopkins University Press | *Baltimore*

© 2014 Johns Hopkins University Press
All rights reserved. Published 2014
Printed in the United States of America on acid-free paper
9 8 7 6 5 4 3 2 1

Johns Hopkins University Press
2715 North Charles Street
Baltimore, Maryland 21218-4363
www.press.jhu.edu

Library of Congress Cataloging-in-Publication Data

Danbom, David B., 1947–
Sod busting : how families made farms on the nineteenth-century plains / David B. Danbom.
pages cm. — (How things worked)
Includes index.
ISBN-13: 978-1-4214-1450-8 (hardcover : alk. paper)
ISBN-13: 978-1-4214-1451-5 (pbk. : alk. paper)
ISBN-13: 978-1-4214-1452-2 (electronic)
ISBN-10: 1-4214-1450-3 (hardcover : alk. paper)
ISBN-10: 1-4214-1451-1 (pbk. : alk. paper)
ISBN-10: 1-4214-1452-X (electronic)
1. Agriculture—Great Plains—History—19th century. 2. Farmers—Great Plains—History—19th century. I. Title. II. Title: How families made farms on the nineteenth-century plains.
S441.D28 2014
630.978—dc23 2013044629

A catalog record for this book is available from the British Library.

Special discounts are available for bulk purchases of this book. For more information, please contact Special Sales at 410-516-6936 or specialsales@press.jhu.edu.

Johns Hopkins University Press uses environmentally friendly book materials, including recycled text paper that is composed of at least 30 percent post-consumer waste, whenever possible.

For Dan

CONTENTS

PREFACE

American history textbooks tell us that homesteaders settled the Great Plains. But none of them explain how that process unfolded. As a part of the series "How Things Worked," this volume explains how white Americans settled the plains region—how newcomers obtained land, built farms, acquired capital, created institutions, founded towns, and did all of the other great and small things that replaced native American tribes and roaming buffalo with farms, fields, and towns.

A huge region, the Great Plains stretch from northeastern Mexico northward into the prairie provinces of Canada. This book focuses mainly on four states—Kansas, Nebraska, and North and South Dakota—that share a continental climate and a similar topography. Each was settled mostly after the Civil War. The railroads assumed a major role in the development of this area, and the Homestead Act and other federal land legislation proved significant. European immigrants played a crucial part in the settlement of these states. Unlike the states on their western borders, they consisted almost entirely of grasslands and did not depend on extractive mining and timber economies. In contrast to Oklahoma and Texas, Kansas, Nebraska, and North and South Dakota were not originally extensions of the Southern cotton culture, with its social, economic, and racial characteristics.

I hope that students reading *Sod Busting* will come to understand that the settlement process on the plains went beyond simply building railroads and confining native people to reservations and that farm building involved much more than simply claiming a piece of land under the Homestead Act and throwing seed on the ground. Building institutions and communities was a long and challenging process, marked by failures as well as successes. Settlement was a complicated process that differed from place to place. That was how things worked.

I would like to thank a number of people for helping make *Sod Busting* possible. Carroll Engelhardt, Sterling Evans, Barbara Handy-Marchello, Paula Nelson, and Claire Strom were kind enough to read my first draft and to provide me with excellent suggestions. Bob Brugger and his staff at Johns Hopkins University Press shepherded this process along through several drafts. Series editors Robin Einhorn and Richard R. John provided several helpful critiques, and two anonymous readers furnished detailed and helpful evaluations. In the early stages of this process, Deborah Sayler, Lorrettax Mindt, and their staff at North Dakota State University's Interlibrary Loan office found the frequently obscure publications I needed. Later that function was undertaken by the Interlibrary Loan staff at the Loveland (Colorado) Public Library.

By this point Karen is an old hand at being married to an author. Still, I appreciate her tolerance of my long silences, pacing about, and hours of isolation in my basement office.

Sod Busting is dedicated to my brother Dan. He is an occasional traveling companion, a frequent fishing buddy, and a constant fellow enthusiast for the Denver Broncos. Most of all, he has been a lifelong supporter of me and my endeavors. There may be better brothers in the world, but you'd have to prove it to me.

Sod Busting

Prologue

SHORTLY AFTER the Louisiana Purchase of 1803, the North West Company, a Canadian firm specializing in animal skins, sent a trader, Alexander Henry, to the Red River Valley in what later became North Dakota. Born in New Jersey and raised in Montreal, Henry could hardly have prepared for what awaited him on the Great Plains. Winter brought snow so deep he had to substitute sled dogs for horses when he visited company posts. He suffered from "snow blindness" and frostbite and once got lost for several days in a severe blizzard. In spring the prairie bloomed, but horses sank to their knees in thick mud, and swarms of mosquitoes plagued man and beast. Driven half mad by the insects, horses "threw their riders . . . or trampled them . . . at night" while they slept on the ground. Henry and his men gained some relief from the relentless bugs by caking their skin with mud and crouching around smoky fires. Late summer and fall brought relief from mosquitoes but heightened the danger of raging prairie fires, which twice burned Henry's trading post.[1]

Nearly a century later Rachel Calof arrived at a homestead a little west of the Red River, where Alexander Henry had tried to make a living. She was a Russian Jew and, by contrast, arrived in what had become the state of North Dakota by steamship and railroad. Yet her experiences closely resembled his. In her first year she found herself lost in tall prairie grass. During the first brutal winter, the Calofs, to save fuel, had to share a shack with two other families, along with a calf and a flock of chickens. The closest water was in a "slough," a marsh a mile away. Users had to be careful to strain it to remove grass and worms. A spring cloudburst once flooded the family's house. One summer a hailstorm destroyed the wheat crop, broke out all of the windows of the house, and killed two horses.[2]

Alexander Henry and Rachel Calof left detailed accounts of their experiences, which are rare. Most Europeans and Americans of European descent arriving on the plains left no accounts at all, appearing only (and many only once) in the federal census, land-transaction records, or property-tax books. But enough memoirs and reminiscences survive to leave no doubt: The plains imposed severe hardship on any and all settlers. We know of the experiences of Mary Dodge Woodward, who coped with the isolation of life on a farm and the cruelties of winter on the northern plains, and of Roderick Cameron and his family, who built a farm in a desolate region of northwestern Kansas, sixty miles from a town of any size.[3] Kansans Catherine Porter and Hattie Lee both lost fathers soon after settling and were thrust into adult responsibilities while still children. Luna Kellie struggled with drought, grasshoppers, and poor crops on a Nebraska homestead, but her greatest burden was avaricious railroads and moneylenders. She responded by becoming a leader in the Farmers Alliance, forerunner of the Populist Party, in the 1880s. On the other side of the political and economic divide, mortgage company employee Seth Humphrey worked to recover his employers' capital from hard-pressed borrowers in South Dakota and Nebraska.[4]

These people differed in a variety of ways, but they were bound by the experience of living, working, seeking success, and building communities in one of the nation's least hospitable climates—the Great Plains of North America, which (in today's geography) stretches from the north-central and northeastern states of Mexico, through the central part of the United States, into the prairie provinces of Canada.

When Europeans and Americans of Europeans descent began settling on the plains, the region already had a long history of human habitation. Paleo-Indian hunters visited the area as early as 17,500 years ago, as the glaciers covering most of the plains receded, and archeologists have found ample evidence of their camps dating to 12,000 years ago. Evidence unearthed thus far indicates that these Paleo-Indians were nomadic hunters who did not establish settled villages on the plains. About two thousand years ago, parts of the region were settled by Mound Builders—so called because of their practice of building earthen mounds for burial and ceremonial purposes—who probably came from the Mississippi Valley.

By the time the first Europeans entered the region, it was at least partially inhabited by settled American Indian groups. When Spaniard Francisco Vasquez de Coronado explored present-day Kansas in 1541, he found people he called "Quivirans"—probably ancestors of modern-day Wichitas or Pawnees—living in agricultural villages along river bottoms. Other American Indian groups had also settled on the plains by that time, such as the Mandan, with whom the Lewis and Clark exploring expedition wintered in 1804–5. The Mandan had relocated from the Mississippi Valley to present-day North Dakota between 1300 and 1400.

Two aspects of this story of early human habitation on the plains are worthy of note. First, the population of the plains was not static. Groups came in and groups went away, replaced—and sometimes displaced—by others. That process continued, and even accelerated, after European contact with the Western Hemisphere began. The result was that some of the Indian groups that came to be associated with the Great Plains were relatively recent migrants to the region. The Cheyenne, for example, came from the Great Lakes region about 1500. And the Sioux migrated from the Upper Mississippi Valley around 1700, driven west by their adversaries, the Ojibwa, who had acquired guns from their French allies.

It is also noteworthy that the Quivirans, the Mandan, and the other American Indian groups occupying the plains in the sixteenth century were not mainly hunters and certainly not the mounted hunters we think about when we contemplate Plains Indian life—they were farmers. The agricultural system practiced by Plains Indian groups was similar to that practiced by most North American Indians at the time European contact began. It was based on three crops—corn, beans, and squash (including pumpkins)—which were planted together in hills and were grown along river bottoms, where fertility was high and moisture was relatively dependable. Women were responsible for planting the crops, harvesting them, and preparing them for winter storage. Corn, beans, and squash had a symbiotic relationship. The corn plant formed a trellis for the beans. The beans drew nitrogen, which was needed by the squash and the corn, from the air and fixed it in the soil. And the squash overspread the ground, inhibiting weed growth and slowing evaporation. These crops were valuable to people who lacked refrigeration or canning technology. Corn and beans could be dried and would last for months when stored in dry pits lined with tree branches or grass. Squash can last a long time, as anyone who has found an ancient zucchini in the back of the refrigerator's vegetable bin can attest, and when dried by means of smoking can be kept for months. These basic crops provided about 80 percent of the caloric intake of Plains Indians, with the rest coming from hunting, fishing, and gathering.

The major change to the American Indian economy and lifestyle on the plains came with the introduction of the horse, a domesticated animal brought to the Western Hemisphere by Europeans. Comanche Indians on the southern plains obtained horses from Pueblo Indians in present-day New Mexico about 1680. Within a century Indian groups throughout the plains had acquired horses through trade, theft, war, or the capture of feral animals. Horses were available to all Plains Indians, but among some of the most powerful groups, such as the Comanche, Cheyenne, and Sioux, their possession initiated a shift from an agricultural economy to a hunting and fur-trading economy.

Other European products also had an effect on Plains Indian life. Material goods, from blankets and cookware to hatchets and firearms, changed daily lives and tempted Indians to involve themselves more heavily in trade. Much less positive was the introduction of European diseases, such as smallpox, that had a substantial negative effect on Indian populations and was especially devastating to people living in settled villages. European contact meant more than new goods and new diseases, however. European nations also claimed and attempted to control the plains. At various times, Great Britain, France, Spain, and Mexico claimed parts of the region. But by the early nineteenth century, the proximity and expansionist inclinations of the United States made it the primary claimant.

The United States acquired title to most of its portion of the Great Plains through the Louisiana Purchase. It gained control over smaller portions under the Convention of 1818, which drew the border between the United States and Canada along the Forty-Ninth Parallel, the annexation of Texas in 1845, and the Mexican Cession in 1848. This region eventually encompassed virtually all of Kansas, Nebraska, Oklahoma, and the Dakotas, along with parts of Texas, Montana, Wyoming, Colorado, and New Mexico.

The purchase of Louisiana prompted the government to sponsor expeditions to explore parts of the region, including the northern and central plains. The Lewis and Clark Expedition of 1804–6 is the most famous of these. Meriwether Lewis, William Clark, and their party ascended the Missouri River from Saint Louis, crossed the Rocky Mountains, and reached the Pacific Ocean at the mouth of the Columbia River. In 1806 and 1807 a party led by Zebulon Pike followed the Arkansas River through present-day Kansas and Colorado, reaching the Rockies. And in 1820 Stephen Long ascended the Platte River through present-day Nebraska and Colorado. These expeditions helped familiarize Americans with the geography, flora, and fauna of the plains and allowed explorers to name geographical features after themselves, but they failed to excite Americans with the idea of settling there. Indeed, they probably had the opposite effect. Stephen Long's characterization of the area of the central plains he visited as the "Great American Desert" was hardly calculated to attract a flood of settlers into the region.

For forty years after the Louisiana Purchase, policymakers viewed the plains primarily as a vast American Indian preserve. From south to north, the region was dominated by the powerful Comanche, Cheyenne, and Sioux, who probably couldn't have been defeated by the U.S. military even if it had been inclined to try. The status of the plains as Indian country was reinforced when the U.S. government relocated tribal groups from the east to the plains as part of a program to remove them from valuable lands coveted by whites. The most infamous removal involved the Five Civilized Tribes of the southeast—the Cherokee, Creek, Chickasaw, Choctaw, and Seminole—relocated to present-day Oklahoma in the 1830s. Further north, Shawnee, Sauk, Fox, and other groups from the Midwest and Great Lakes were relocated at the same time to what is now eastern Kansas and Nebraska.

Most of the few Europeans and Americans who could be found on the plains prior to the 1840s were fur traders. Many of these men were from fur-trading French, Spanish, and British families, though they lived under American sovereignty and were usually American citizens. These traders frequently married Indian women, sometimes to cement a business relationship or create a political or economic alliance. William Bent, for example, who ran Bent's Fort in what is now Colorado with his brother Charles, married Owl Woman, daughter of White Thunder, a powerful Cheyenne chief. These business relationships were important because most of the gathering and processing of furs on the plains was done by Indians or Métis—people of mixed European and Indian descent. They traded their furs at posts such as Bent's Fort and Fort Union, in present-day North Dakota. In return they received manufactured goods. The forts were supplied and the furs were carried away by steamboats, bull boats, or ox carts. Fur traders took most types of furs, but bison were the most numerous large animals on the plains and the trade in bison skins was the most active. Indians on horseback could kill many more bison than they could use, and the lure of trade goods was strong. As a result of overhunting and the fur trade, the bison population began to decline from its 1820 peak of twenty million even before white buffalo hunters ventured onto the plains.

American interest in the region—and conflicts with Indian groups there—accelerated in the 1840s because of the acquisition of Texas and travel on wagon trails by thousands of settlers bound for Oregon, Utah, California, and later Colorado. The government had earlier negotiated trade agreements and peace treaties with many of the Indians, and it now secured passage rights for emigrants. Movement of settlers across the plains seems minimally invasive on the surface, but it sparked conflicts between travelers and Indian bands and spread disease among the tribes, and the way stations created to serve emigrants shut Indians off from areas along river bottoms where they wintered and secured forage for their horses. In combination with the depletion of the bison herds, emigrant travel along Western trails contributed to the decline of the Plains Indians, setting them on the road to military defeat and confinement on reservations after the Civil War.[5] Midwestern American travelers' increasing familiarity with the plains did not change popular attitudes toward the region. Today's "flyover country" was rideover, driveover, or walkover country in the 1840s—a barrier separating ambitious immigrants from the rich valleys of Oregon, the Mormon refuge of Utah, or the goldfields of California and Colorado.

Antebellum Americans shunned the plains in part because what they knew about the area repelled them. The plains region as a whole does not lack topographical variety. There are ranges of hills and even mountains—such as the Flint Hills of Kansas, the Black Hills of South Dakota, and the Turtle Mountains of North Dakota, as well as badlands in the western Dakotas, and areas that fall somewhere in between, such as the Nebraska Sandhills. But in general the plains consists of natural grassland, ranging from tall grass prairie in the east—where the grass sometimes grew as tall as a horse's back—to short grass in the west, the difference attributable mainly to variations in precipitation. Gently rolling hills, a product of glaciations, characterize most of the region, but to many observers its monotony meant, in the words of one visitor to western Nebraska, that "no more depressing landscape can be imagined." The lack of landmarks, the "treeless plain [that] stretches away to the horizon line," created an effect that, if not otherworldly, could certainly be disorienting for people from places with more varied topographies. Visitors to the plains, seeing nothing but waving grass on every side, most often compared the experience to being at sea. In addition to being disorienting, it could be dangerous. Without the benefit of landmarks, inexperienced people lost their way, as did Laura and Carrie Ingalls, who were lost for hours in a tall-grass slough. Even an experienced plainsman such as Seth Humphrey, whose job as a mortgage company representative compelled him to travel widely in South Dakota and Nebraska, could become confused. "On a cloudy day or a starless night . . . one's sense of direction might take a half turn. Then north would become west or east or south." Humphrey prudently carried a compass at all times.[6]

The monotony of the plains and the absence of landmarks were directly related to the lack of trees. Trees were scarce on the plains in part because of the climate of the region and in part because of the American Indian practice of burning the prairies to enhance the environment for game. As a result, the trees that existed were mainly on the banks of the relatively few rivers. The treeless landscape seemed strange to Americans from forested regions, but it also presented practical difficulties. Nineteenth-century American farmers built and heated their homes and built their fences with wood. Without locally available timber, what would they do?

Water was scarce, and the paucity of surface water offered one indication of a difficult climate. Few rivers and streams flowed through the region, especially in its western reaches, and many flowed only part of the year. To potential farmers from the Ohio and Mississippi Valleys, the lack of easily accessible, navigable rivers meant that inexpensive marketing of bulky crops would be difficult, if not impossible. Potential farmers had to think of providing water for animals, in addition to themselves. Most of the plains region receives an average of less than twenty inches of moisture per year, which mid-nineteenth-century farmers considered the minimum necessary for successful agriculture, and in parts of the western plains the average is barely ten inches. The bulk of the plains region has a continental climate, which means that it is too far from a major body of water—such as the Pacific Ocean or the Gulf of Mexico—to receive abundant and predictable moisture from year to year. Major bodies of water also have the effect of moderating temperatures.

Continental climates tend to the extremes—fearsome highs in summer and lows in winter. North Dakota in 1936 supplied a telling example of the temperature swings incidental to a continental climate: in February the town of Parshall recorded a low of −60 degrees Fahrenheit, and five months later Steele had a high temperature of +121 degrees, for a 181-degree swing.[7] Settlers may have adjusted to these unpleasant realities, along with the others, over time, but the climate would always be inhospitable.

In spite of all the warning signs, developments during the 1850s and 1860s conspired to encourage white settlement on the plains. Government played a crucial role in this venture, and it continued to do so throughout the settlement period and beyond. In 1854 Congress passed the Kansas-Nebraska Act, creating two territories in the central plains and opening them for settlement. Illinois Senator Stephen Douglas sponsored this legislation in an effort to strengthen the case for a central route for the proposed transcontinental railroad. The Missouri Compromise of 1820 had prohibited slavery in this region. To attract Southern support for his bill, Douglas decided to support repeal of the Missouri Compromise throughout the West, leaving the question of slavery up to the settlers—"popular sovereignty," as he and his allies called it. Pro- and anti-slavery forces soon rushed into and battled in Kansas, making "Bleeding Kansas" a notorious rallying cry for both sides. But while the Kansas-Nebraska Act played a critical role in the sectional crisis, it also led to the first significant permanent white settlements on the eastern plains.

Once eleven of the slave states left the Union, congressional Republicans in the crucial years of 1861 and 1862 laid much of the policy foundation encouraging further plains settlement. The absence of Southern Democratic representatives and senators in Congress gave Republicans majorities in both houses. They used their power to pass legislation embodying their belief that a positive government could and should help create a strong, wealthy, and egalitarian nation. In 1861 Congress admitted Kansas to the Union as a free state and organized Dakota Territory as a free territory. Those actions provided reassurance to potential settlers who opposed slavery or, less positively, did not want African Americans living in their midst that the plains would be a land of free labor. In 1862 Congress passed the Homestead Act, which allowed any head of a family, male or female, citizen or alien who had stated an intention to become a citizen, to claim up to 160 acres of land in return for a fairly nominal filing fee. If the homesteader made certain stipulated improvements on the claim, he or she could receive clear title to the land after five years, upon payment of another fee. The Homestead Act had many flaws, applied to much less than half of the land on the plains, and did not by itself result in settlement of the region. Along with its amendments and elaborations, however, it encouraged plains settlement, and it illustrated the government's determination to accelerate western development and expand economic opportunity.[8]

Congressional passage of the Pacific Railway Act in 1862 underscored the Republican commitment to economic opportunity. The Pacific Railway Act was the first in a series of legislative enactments providing loans, land grants, or a combination of both to railroads traversing the plains. Railroads were crucial to substantial settlement because there was no other practical way to move high-volume, low-value commodities cheaply. Moreover, railroad companies played a critical role in promoting the plains, and their desire quickly to sell the lands they received speeded the pace of settlement. Critics attacked the policy of lavishing benefits on railroads, but the fact remains that plains development depended on them and that without government aid the railroads would not have expanded as rapidly and ambitiously as they did.

Less spectacular legislation also helped to lay the foundation for successful settlement of the plains. In 1862, shortly after Congress created the U.S. Department of Agriculture, it passed the Morrill Land-Grant College Act, which gave the states federal land on condition that they use the proceeds of its sale to fund agricultural colleges. Eventually, these pieces of legislation created a scientific and technical establishment that addressed the special problems of agriculture on the plains.

The year 1862 also marked the outbreak of a series of wars in the West between Plains Indian groups and the U.S. Army. These wars, which continued sporadically for the better part of fifteen years, along with the destruction of the bulk of the bison on which Plains Natives depended, resulted in the end of many tribal land claims and the confinement of most Indians to reservations, continuing a process that had begun a half-century before. As a consequence, the government could open vast tracts to settlement by white Americans and European immigrants.

Legislative acts and Indian wars raised the profile of the federal government on the plains, but the scope of government activism was much greater than the sum of its parts. The federal government emerged from the Civil War with a robust conception of its powers and potential. It not only controlled land policy and Indian affairs but also influenced and sometimes directed transportation development, commerce, industrial expansion, resource exploitation, and banking and finance, all of which affected the people of the plains. It is perhaps an overstatement to say that "big government won the West." But the West would have looked very different a century ago, and would look very different today, had development depended solely on private enterprise and initiative.[9] Government was, and remained, a necessary facilitator of plains settlement, but it was not sufficient. Government could give land away, teach farmers to grow crops that were suited to the land and climate, and subsidize railroads to carry crops to market, but those actions would be unavailing in the absence of people willing to come to the plains.

Demographic developments in Europe produced many potential settlers for the plains. In the century after 1750, the European population doubled. Improvements in medicine and sanitation partly spurred that increase, but it was due primarily to the fact that Europeans were better fed. Europeans ate more and better food mainly because of changes in farming practices and land use that increased production per acre. Still, food producers struggled to keep pace with population growth. Western and central European states responded by increasing imports of food, first from regions on the European periphery, such as the United States, Canada, and Russia, and later from more distant countries, such as Argentina, Australia, and South Africa. All of these places became more accessible because of the expansion of railroads that brought produce to ports and steamships that moved it efficiently to consumers. Competition from farmers outside the continent intensified pressures on small landowners, tenants, and laborers in western and central Europe, who already struggled to make a living on small plots. Some of these people left the land for towns and cities in Germany, Great Britain, and other European countries that were industrializing rapidly. But large numbers of rural people believed that emigration offered them opportunities unavailable at home. Their home countries increasingly agreed. After years of discouraging emigration because it drained off their young people, European governments came to believe that they would be better served if their discontented and underemployed people went somewhere else. Those who wanted to do so took advantage of the same transportation facilities that had flooded their countries with cheap agricultural produce from abroad, making emigration decreasingly expensive.[10]

Immigrants went where economic opportunities were greatest, and in the late nineteenth century that frequently meant the United States. Three-fifths of the fifty-five million Europeans who left their home countries between 1850 and 1914 came to the United States. Their presence helped swell a population that was already rising rapidly through natural increase. Between 1820 and 1870 the population of the United States increased by 315 percent, and it rose by another 130 percent by 1910. Immigrants who found work in mines, mills, and factories played a critical role in fueling the phenomenal industrial expansion that made the United States the leading industrial country in the world by 1890. But immigrants also found a country with a vast expanse of free or inexpensive land, increasingly accessible by rail, which offered opportunities not readily available in their home countries. Not all immigrants desired to farm and some of those who did have that desire lacked the resources to act on it. But those who had the will and the means had the opportunity for commercial success in an industrializing Western world needing to be fed. They also had the chance to achieve independence and even to acquire wealth not easily imagined in their home countries. Whatever the risks, these opportunities drew many immigrants to the plains. There they were joined by mobile Americans, mostly farmers who had caught the western fever, as had their fathers and grandfathers, including tens of thousands of Civil War veterans eager for new starts in new places.[11]

There was romance to this westward movement, and it remains embedded in some of our most enduring myths. Yet few of these plains settlers were romantics. They were modern men and women, settling a frontier that benefited from modern transportation and an active and supportive government, using mass-produced agricultural implements they hoped would allow them to exploit the market opportunities offered by an industrializing world and build their wealth. They needed to be resourceful, energetic, and sometimes courageous, but they were not rugged individualists. They depended on opportunities provided by government, and they were part of a modern world that sold them implements and household goods; stored, shipped, and processed their crops; and leant them money. A crucial part of the settlers' story is how they got capital, used labor, interacted with the transportation, storage, and processing facilities supporting agricultural production, and created the social institutions that gave meaning to their lives. It all began with acquiring land.

1 How They Acquired Land

LAND IS BASIC to agricultural production, and acquiring it was the first necessity for settlers on the Great Plains. Most of the land on the plains belonged either to the federal government or to Native American bands and tribes with which the government had made treaties. In the early 1850s the government began the process of reducing American Indian holdings on the eastern plains by extinguishing land titles and writing new treaties. Sometimes Indians were forced to relinquish territory as a result of war, most notably at the end of the Civil War, when the Five Civilized Tribes were compelled to surrender about half of their land in Oklahoma as punishment for supporting the Confederacy. More frequently, Indians surrendered land to settle trading debts, to receive more liberal and dependable annuity payments, or because they were promised greater security from encroaching settlers. This process was supposed to operate smoothly and systematically, but in practice it seldom did. The government was sometimes slow to make treaties or picked emissaries who were incompetent or inefficient, and finding Indians in leadership positions who were willing to negotiate on terms attractive to federal representatives was not always easy. Then there was the problem of Congress, where the Senate had to approve treaties and the House had to appropriate the funds that were usually required to fulfill the government's part of the deal. Sometimes the process broke down entirely, as in Kansas, where the territory was declared open for settlement before any tribal land titles had been extinguished. The resulting confusion dramatically increased the conflict and disorder inevitable in a territory that Congress had turned into the primary focus of the sectional conflict over slavery.[1]

The process of rewriting treaties and relocating Native Americans accelerated after the Civil War and shifted to the central and western plains. Government progress was slower in that region, and Indian groups were more likely to offer armed resistance. Some Comanche, Cheyenne, and Sioux bands were highly mobile, militarily skilled, and unwilling to surrender tribal lands without a fight. Eventually the army got the upper hand, in part because of the institution of winter campaigns and in part because of the decimation of the bison herds on which many Indians depended. By 1880 the major Plains Indian groups had surrendered to federal authorities and were confined to reservations. In many cases American Indians probably made the best bargains possible under adverse circumstances, but the government did not consistently fulfill its treaty obligations. In 1876, for example, when European American prospectors swarmed into the Black Hills, part of the Lakota Sioux reservation, the military ended up protecting the settlers after a half-hearted attempt to uphold Native American treaty rights. Even when Indians managed to retain title to reservation lands, their hold was seldom secure. White cattlemen frequently grazed their cattle on reservation land, sometimes with the approval of Indian agents or tribal leaders and sometimes without. As white settlement increasingly encroached on tribal lands, the government again pressured the Indians to cede portions of reservations. The new mechanism for accomplishing this goal was "severalty"—the end of tribal land ownership through the transfer of property to individual Indian families.[2]

The concept of severalty was embodied in the Dawes Severalty Act, passed in 1887 and subsequently amended several times. The stated purpose of this legislation was to speed assimilation of American Indians by inducing them to take up individual land allotments of 160 acres per family. On these allotments they were expected to become farmers in the white American style and to develop individualistic values that would erode their tribal allegiances. To advance the planned transformation of the Indians, the government promised agricultural instruction and equipment, and to protect them from rapacious whites the government retained title to their farms for twenty-five years.

Severalty was a failure in almost every way. Agricultural instruction was often indifferent when it was offered at all. Promised equipment frequently did not appear, and when it did it was often second-rate. Annuities promised to the tribes that would have helped families survive on small farms were often shaved by penurious Congresses or siphoned off by corrupt Indian agents and contractors. Under the best of circumstances, 160 acres of land in the arid West was hardly sufficient for agricultural success; when it was divided by heirs of original owners, unlikely success became impossible. The most significant problem with severalty was that it was totally alien to Indian culture. People with strong communal values were required to act as individualists, and men who were hunter-warriors were expected to be farmers, an occupation traditionally held by women. This was a recipe for disaster.

There was one way, however, in which severalty succeeded—it severed the tribes from much of their remaining land base. Once lands had been allotted to families and individuals, the government declared what remained to be surplus. Tribes were paid for these surplus lands, which were then opened to white settlement. By 1900 the Plains tribes had lost nearly half of their reservation land through this process. They were compensated, but the low prices they received did not begin to meet the long-term cultural and economic costs they sustained due to the erosion of their territory. Severalty ended up being one of the sadder chapters in the dismal history of government-Indian relations on the plains.

Once Indian holdings were transferred to the federal government, it began the process of preparing the land for conveyance to settlers or to various political and corporate entities by surveying it. The methods of survey had been laid out in the Basic Land Ordinance, passed by Congress in 1785. The Basic Land Ordinance provided that federal land be divided into square "townships" six miles on a side. Each of the townships was further divided into 36 "sections" each one mile square, or 640 acres. The Basic Land Ordinance reflected an eighteenth-century passion for rational order and a preference for systematic settlement and land disposal. At the same time, it ignored the natural features and contours of the landscape and has been condemned for encouraging people to think of the land as a commodity. The advantages of the rigid survey were greater than these shortcomings. The United States embraced a system of private property and agricultural capitalism from the beginning, and farmers would have viewed land as a commodity regardless of how it was demarcated. At least under the Basic Land Ordinance parcels were clearly and consistently identified, allowing settlers to avoid the extended and acrimonious land title disputes that bedeviled parts of the Atlantic seaboard and the South, where the standard survey was not applied. On the plains the survey created landmarks where precious few existed. Section lines, along which fences and roads frequently ran, served to orient people and help them locate their position in the absence of trees, streams, or ridges. An airplane passenger flying over the plains today can see the resulting orderly demarcation of what had been a vast and relatively featureless landscape.

When the land was surveyed, it could be legally acquired by settlers, but from whom? The government maintained control of much of the land. It conveyed some to settlers through the Homestead Act, the Timber Culture Act, and other pieces of legislation and sold some, usually at auction. But most of the land went to the states when they entered the Union or chartered land-grant colleges or was granted to railroads. For the settlers, acquiring land was a complicated process of choices. The settler could homestead virtually for free, buy federal land, or buy land from a state, a railroad corporation, a land company, or, increasingly, another settler. Settlers often acquired several parcels of land in several ways. This chapter explores the available choices, the complications and benefits of each, and the decision-making processes settlers followed.

Federal Government Land Policies

Of all the areas of federal endeavor in the first 125 years of our national existence, none was as complicated or contradictory as land policy.[3] In the early years of the republic, when the government operated under the Articles of Confederation, land was the only reliable source of revenue for Congress, leading it to undertake large-scale sales to speculators. The idea that the government should profit from the disposal of land never totally disappeared, but over time, as democratic ideals rose to greater prominence in political life and thought, Congress made land more readily available to settlers by limiting its price, allowing smaller parcels to be bought, and permitting buyers to purchase it on credit. The government also enabled squatters—people who settled on property they did not own—to purchase under reasonable terms the land on which they had settled. At the same time, the government used land to reward veterans, provide income for new states, and encourage infrastructure development and the creation of useful public institutions. The railroad land grants and the land grants for colleges of the Civil War era, though remarkable for their size, were well within this tradition.

By the 1840s some in Congress, such as Stephen Douglas of Illinois and other Democrats from the Midwest, had embraced the idea that parcels of public land should be given to settlers more or less for free. The new Republican Party rewarded former Democrats in its ranks by including the free land proposal in its 1856 and 1860 platforms, and it was achieved with the passage of the Homestead Act in 1862. By that time, Republicans had come to see this policy not merely as an act of political expediency but also as a means of enriching the nation by rapidly broadening its economic base. The Homestead Act, which played a significant role in encouraging settlement of the plains, allowed any head of household at least twenty-one years of age to acquire up to 160 acres of public land. Homesteaders could claim smaller parcels, but seldom did, for the simple reason that 160 acres of free land is preferable to 80 acres of free land. Women who were heads of households, such as those who were single or widowed, were allowed to claim homesteads, as were immigrants who had taken out "first papers" stating their intention of becoming citizens. Former slaves were also allowed to claim land, though few did because of their lack of capital and because of the hostility toward them among Western settlers. The settler was required to file a claim, pay a fee, establish residency on the land, and improve it in certain stipulated ways. At the end of five years, the settler could "prove up"—demonstrate that he or she had met the government's requirements—and, after paying another fee, receive title to the land. It all sounded so simple, but in practice the process was devilishly complicated.[4]

For one thing, there were exceptions to what was, on the surface, a remarkably liberal and permissive piece of legislation. Participants in the Rebellion, including veterans of the Confederate Army, were not allowed to file. The Civil War Congress was in no mood to reward people it regarded as traitors. Moreover, a filer could legally take only one homestead of 160 acres over his or her lifetime. Failure on a homestead claim did not prevent a farmer from starting over somewhere else, but he or she could not legally start over on free government land. In practice, though, this prohibition was widely violated by settlers who changed their identities, a fairly simple process before the creation of Social Security, when Americans received what amounted to national identity cards. As Seth Humphrey explained it, "Bill Jones of the Wisconsin boom thought of himself as Hank Brown in Minnesota, then shifted to John Smith for his filing in Dakota."[5]

Even qualified settlers expecting to find vast expanses of land open to homesteading were frequently disappointed. Of the 500 million acres of public land transferred to private hands between 1860 and 1900, only about 80 million—or less than one-sixth—was available under the Homestead Act. Another 300 million acres were granted to the states or to the railroads. About 110 million acres was sold by the government, usually alternate sections within railroad land grants. Nor were all homestead entries exactly free. When the government opened reservation lands for homesteading, for example, it charged settlers $1.25 to $6.00 an acre to compensate Indians whose lands were surrendered.[6]

These were not the only limitations with which homesteading settlers coped. Homesteaders could claim government land adjacent to railroad land grants, but unless they were Union veterans, for whom many exceptions were made, they could claim only eighty acres. This limitation was designed to encourage settlers to buy railroad land, which would contribute to the financial success of the railroads.

Even the land that seemed to be available wasn't always available. Homesteaders could not file on the land until it was surveyed, but squatters could settle on it at any time. Under the Pre-Emption Act of 1841, a settler squatting on the land for fourteen months and making improvements could buy up to 160 acres for $1.25 an acre. It was not uncommon for a settler to acquire one quarter section under the Pre-Emption Act and then claim an adjoining quarter under the Homestead Act. There were also problems with land that was claimed but poorly marked, land that had been claimed but abandoned by the claimant, and land on which the claim had been transferred by the original claimant to another. In this difficult and confusing situation, prospective homesteaders turned to professional "locators" or "land scouts." For a fee that usually ranged from ten to thirty dollars, a land scout would guide a new settler to land that was unclaimed, unowned, and agriculturally promising. Land scouts were costly, but the expense was justified by the professional services they offered. Old Jules Sandoz charged twenty-five dollars for scouting land in the Nebraska Sand Hills, but his services included accompanying the settler to the land office in Alliance, helping him or her complete the paperwork on the claim, and surveying the homestead when the settler wanted to fence. Those lacking the cash could work it off by breaking land on Old Jules's farm or providing him with firewood.[7]

Homesteaders frequently sought adjoining homesteads for friends, fiancés, or relatives. Sometimes settlers were seeking the comfort and security provided by close neighbors. In 1913 author Mabel Stuart wrote of two sisters and a brother in Butte County, South Dakota, who clustered their homes together where the corners of three quarter sections met. Their little neighborhood was soon expanded when two sisters built on adjoining claims. Within days of taking out her first naturalization papers in 1903, Norwegian immigrant Mina Westbye and two cousins filed on neighboring quarters in Divide County, North Dakota. While commentators emphasized the tendency of women such as Westbye and her cousins to settle together, the practice was also common among men. Relatives or engaged couples who settled on adjacent parcels were able to expand family holdings while enjoying the companionship of nearby neighbors. Rachel Kahn, betrothed to Abraham Calof, took out citizenship papers and claimed a quarter section next to his. And Elinore Pruitt Stewart, famous for her "Letters of a Woman Homesteader," took up a claim next to those held by her husband-to-be and her widowed mother-in-law-to-be. Such strategies for family accumulation of land were all within the letter of the law but had probably not been foreseen by its authors.[8]

Once a homesteader had identified his or her land, marked it with corner stakes, a "straddle-bug" (tripod made of boards), or some other contrivance, and filed the claim, he or she set about fulfilling the requirements of the Homestead Act. A homesteader had a six-month grace period before taking up residence and could prove up in a minimum of five years or a maximum of seven years. Those who waited seven years usually did so because of their difficulty in meeting the act's stipulations or because they wanted to delay receiving title and having to pay property taxes. The homesteader was required to erect a dwelling, establish residency on the claim, break a portion of the land, and commence agriculture.[9]

It sounds simple enough, but parts of the process—especially the residency requirement—could be confusing. Homesteaders were required to live on their land for five years, but what exactly did that mean? If Congress meant that they had to be present for 365 days every year, few homesteaders would meet the requirement. It was common for single homesteaders, especially, to leave their claims in the winter. Some went to distant sites to work and others took jobs in nearby towns. Was one a "resident" when he or she lived on the claim in the summer and on the occasional weekend or holiday? The residency requirement was usually interpreted permissively, but there was always the danger that a homesteader's right to a claim could be challenged by a neighbor who seldom saw him or her around and who coveted the land.

Congress further complicated the process by changing the rules periodically or by writing new rules, especially on behalf of veterans of the Union Army, who were allowed to apply their years of service to the five-year requirement. Widows of Union soldiers were given the right to apply their husbands' years of service against the five-year requirement. All widows inherited their husbands' claims and had the right to complete the homestead process.

Once five years had passed and the homesteader believed he or she had fulfilled the requirements of the act, it was time to "prove up." The claimant was required to file an affidavit detailing "the improvements he or she had made to the property." In addition, the settler and two witnesses had to affirm that he or she had no loans secured by the land, that the homesteader had fulfilled the residency requirement, "that the settler had remained loyal to the United States, and that the settler was or had become a full citizen of the United States." This last step was fraught with danger. Land office employees could disqualify an application because of fraud or because the applicant had failed to fulfill a requirement. Those attempting to prove up were sometimes challenged by rival claimants, frequently aided by "actual or would-be lawyers . . . contest sharks," as Mari Sandoz called them, especially for failing to fulfill the residency requirement. Factor in the inexperience of the politically appointed land-office employees, their corruption, and the complexity of the laws they were called on to administer, and the hazards of the process become abundantly clear.[10]

If the proving-up process was negotiated successfully, the claimant got a receipt from the land office. His or her deed to the land did not come until the patent application had been reviewed and approved in Washington, a procedure that usually took about a year. Denials of applications at this final level were rare, so approval by the local land office made the claimant the de facto owner of the land. When he or she filed the land office receipt at the county courthouse, the land could be mortgaged.[11]

Despite the attraction of the idea of free land, only a minority of homestead entrants proved up, and that minority diminished over time. Some homesteaders simply abandoned their claims, but it was more common for settlers to short-circuit the lengthy homesteading process through "commutation" or "relinquishment." Commutation involved purchasing a homestead claim instead of proving up. The Homestead Act stipulated that any entrant could commute his or her claim and receive title to the land after six month's residence (fourteen months after 1891) for $1.25 per acre, provided he or she had broken at least one acre. Commutation appealed to homesteaders who were unable or unwilling to fulfill some of the requirements for proving up. For example, immigrants who did not desire to become citizens could gain title to their claims through commutation. Commutation was also an attractive option for settlers who wanted to sell the land they had claimed. Nina Westbye, who had difficulty fulfilling the residency requirement on her North Dakota claim, commuted it in 1906 for two hundred dollars. This allowed her to sell it, which she did two years later for one thousand dollars. Settlers who wanted to borrow also found commutation attractive. Some lenders would extend credit to homesteaders, but the loan could not be secured legally by the claim. Owners held secure collateral and could expect to receive larger loans at more favorable terms. Thor and Gjertru Birkel, Norwegian immigrants living in western North Dakota, needed to replace a house gutted by fire. A local lender agreed to advance them six hundred dollars, but only if he could take a mortgage on their land. For them, commutation was a necessity. For most it was an attractive economic strategy.[12]

Commutation allowed a homesteader to acquire title to his or her land quickly. Relinquishment allowed a homesteader to realize some return from his or her claim by transferring it to another. A homesteader relinquishing a claim made an arrangement to surrender it to another claimant. In return for surrendering his or her claim, the relinquisher usually received a financial consideration that could range from fifty to several hundred dollars. It was also common for relatives to relinquish claims to one another. In those cases, money usually did not change hands. Because the Homestead Act did not provide for the transfer of claims, the relinquisher and the new claimant would go to the land office together. The owner of the claim would formally surrender it, and the new claimant would step forward to take it. Relinquishment was a device favored by land speculators, and it was not uncommon for the same quarter section to be relinquished several times before it finally passed to an actual settler.

Relinquishment contributed to the atmosphere of impermanence and flux that pervaded homesteading areas. Rapid and frequent turnover of land seemed more indicative of places where people speculated than places where they built homes. Seth Humphrey argued that "by far the greater number of landseekers took up government land with the intention of unloading it on somebody else." Contributing to this perceived problem was the supposed nature of people attracted by government land that could be claimed for a nominal fee and sold quickly for a profit. Settlers on Kansas homestead lands, for example, were characterized as "the defeated, the failures, and the unsuccessful." Regarding homesteaders in Nebraska and South Dakota, Humphrey concluded that "bona fide farmers" were "thrown together with ne'er-do-wells . . . renters tired of renting, others merely tired of the places they were leaving, and . . . pseudo-settlers . . . restless clerks, tired professors, schoolma'ams, and the like."[13]

Complaints about relinquishments and the homestead claimants who used them intensified after 1900, when annual homestead entries were more than twice what they had been prior to 1900. The explosion in homestead claims when the western plains were being opened to white settlement was attributable to four main factors: more permissive land laws, especially regarding residency requirements; the extension of railroads into hitherto unserved regions; rising land values nationally and regionally; and the attractiveness of homesteading to single men and, especially, women, who sometimes had little or no agricultural experience.

These young, single men and women were frequently stigmatized as "speculative homesteaders" in search of a quick profit and were contrasted unfavorably to "bona fide" settlers committed to building homes and communities. That characterization is simplistic and unfair. Many young female homesteaders took up claims to advance their independence. For Mina Westbye, whose profit from the sale of her homestead allowed her to receive training to become a professional photographer, land ownership was the means to independence. After she proved up on her claim in Kit Carson County, Colorado, Alice Newberry rented out the land and moved to Denver to teach school. Income from her claim supplemented her meager salary and later her retirement. For others, the homestead was an end in itself. As one contemporary magazine author noted of a woman homesteader she had interviewed, "Would she go back? Well, she should guess not! . . . No more submitting to a dictating father and brother, who think they know it all . . . It's better . . . under your roof than theirs." Nor was it fair to condemn young homesteaders for their supposed economic motives at a time when quarter sections on the western plains could be sold for four hundred to twenty-six hundred dollars. Virtually all plains settlers in the capitalist economic system were speculators in one sense or another; all anticipated that rising land values would enhance their net worth and prosperity in either the short or the long term. While those who held land and built farms were praised for being "bona fide" settlers in contrast to relinquishers, rent-seekers, and those looking to sell quickly, all were pursuing their self-interest in a system that encouraged individuals to do so.[14]

The Homestead Act was a watershed in federal land policy, but it was hardly the culmination of land legislation. Congress tinkered with the ramshackle structure in almost every session, adding new provisions and modifying old ones. New laws provided incentives to settlers to irrigate dry lands or drain wet ones. In 1912 Congress modified the Homestead Act by allowing claimants to prove up in three years and lowering the residency requirement to seven months per year. In 1913, Congress passed the Enlarged Homestead Act, which permitted settlers in Montana, Wyoming, Colorado, Utah, Arizona, Nevada, and New Mexico to claim 320 acres of land—twice the maximum allowed under the original Homestead Act. But the most important elaboration of the Homestead Act for people on the Great Plains was the Timber Culture Act, passed in 1873, which allowed a settler to claim an additional 160 acres of land and to receive title to it if he or she planted 40 of the acquired acres in trees.

Mixed motives accounted for passage of the Timber Culture Act. Some of its supporters believed that 160 acres were insufficient to support a family on the forbidding plains and sought to help settlers expand their farms. Others saw the growth of trees as a reasonable way to address the shortage of wood for fuel, fencing, and housing on the plains. Then there was the imaginative argument that trees increased rainfall by drawing moisture from the earth, which then evaporated from the leaves into the atmosphere. This idea may sound dubious to us, but it was one of several theories suggesting that humans could change the plains climate; these theories were popular in the late nineteenth century, even among reputable scientists.[15]

The requirement in the original act that forty acres be planted in trees—inserted by congressmen far removed from any actual physical labor—was so onerous as to be ridiculous, and it was replaced in 1878 by a requirement

Homesteader Randi Garmann washing clothes outside her claim shack, Williams County, North Dakota, about 1910. She has banked part of her house with sod in preparation for winter. Institute for Regional Studies, North Dakota State University, Fargo (2008 116.61). Used by permission.

that claimants have a minimum of 675 living trees on each of at least ten acres. Even at this reduced level the tree-planting requirement was daunting, but the other requirements of the act were quite permissive. Upon payment of a small fee, the claimant was allowed to use the land for up to thirteen years and, because he or she did not have title, no taxes were assessed. The claimant did not have to plant trees on the land or improve it in any other way unless he or she intended to prove up. The vast majority of Timber Culture claimants apparently had no such intention. They intended instead to hold the land temporarily for themselves or for family members or, more frequently, to sell relinquishments, which were even more common with Timber Culture claims than with Homestead claims. The combination of unrealistic requirements and speculators adept at gaming the system resulted in widespread frustration of Congress's intentions in passing this act and in broad circumvention of its provisions, with nearly 90 percent of deeds issued under the Timber Culture Act failing to satisfy the letter of the law.[16]

Relatively few settlers were satisfied with holding a mere quarter section, and through some combination of homesteading, making timber claims, pre-empting land, and purchase, they were able to put larger farms together. When Roderick Cameron's father moved the family to northwest Kansas in 1878, for example, he homesteaded 80 acres, as much as he legally could because he had earlier homesteaded 80 in Minnesota. He put Roderick and another son on quarter sections as pre-emptors; neither was twenty-one yet, but they could hold the land until they came of age and could legally take title. He pre-empted another 160 acres for himself and filed a timber claim on still another quarter section. Through these maneuvers the Camerons were able to create a 720-acre farm. Such aggrandizement was generally regarded as acceptable at the time, and still is, because 160 acres was insufficient for agricultural success in the region. Larger holdings were certainly necessary in the central and western plains, where the Camerons settled, and where low rainfall made dependence on crop production problematical. But on the eastern plains, 160 acres was as much as a family could reasonably expect to farm, given the available agricultural technology. That settlers were so eager to accumulate more was undoubtedly due to several factors, including the desire to provide land for their children and other heirs, but a significant consideration was unquestionably the opportunity to acquire resources at a fraction of their anticipated value. It was not merely small-scale settlers who took advantage of the land laws to expand their holdings. Land syndicates, substantial cattle ranchers, and others made full use of pre-emption, commutation, and relinquishment to gain control of large parcels. Sometimes they used "dummy entrymen," employees or temporary hirelings who filed or squatted on land with the understanding that they would turn it over to their employer at the first legal opportunity. While these grandees did not usually violate the letter of the law, they offended its spirit. In the process, they engendered resentment from their neighbors with smaller holdings and prevented the West from becoming the small farmers' republic some had envisioned.[17]

Purchasing Land

Most of the land on the Great Plains—perhaps as much as 80 percent—was purchased rather than acquired under the Homestead Act. About 300 million acres was transferred to the states and the railroads, which in turn sold some to speculators and some to settlers. Over 100 million more acres were held by the federal government and were not opened to homesteading. This land was auctioned periodically, at a minimum price of $1.25 an acre outside railroad land grants and $2.50 an acre within.[18]

Why would people purchase federal land when the government was giving it away? The main reason lies in the key factor in the real estate business—location, location, location. Purchasers especially coveted federal land within a railroad land grant because of its proximity to transportation. In the first couple of decades after the Civil War, before the railroads began extensive development of branch lines, homesteaders were usually twenty to fifty miles distant from the nearest railroad. Because a team of horses pulling a load could travel no more than twenty miles per day under optimal conditions, this meant one to three days in and one to three days back to deliver a load of grain. It was largely for this reason that lands close to a railroad were so desirable and commanded a premium price—nearly 25 percent more per acre than land distant from a railroad. The other reason for purchasing government land was that it was easier to put large parcels together. This attracted large ranchers and land companies to the government auctions.[19]

Purchasers of government land, especially those who bought large blocks, seldom paid cash. Instead, they used warrants or "scrip." Warrants were authorizations to receive federal land in return for services rendered, most commonly military service. Most of these were sold by the original recipients and could be used by anyone. Scrip was a certificate representing a certain acreage of land at the minimum price of $1.25 per acre. The government gave it to Indian tribes surrendering some of their lands in exchange for other lands. Scrip was also given to states creating agricultural colleges under the Morrill Land-Grant College Act. Under the terms of that legislation, each state received scrip representing thirty thousand acres of federal land for each of its U.S. representatives and senators. It could exchange the scrip for federal land, if it had any within its borders, or it could convey the scrip to an existing college that agreed to offer courses in agriculture and the mechanic arts, as stipulated by the Morrill Act. But if it was looking for cash to fund a new agricultural college—which most states were—it simply sold its scrip, usually for forty to seventy cents an acre. In addition, buyers of federal land were allowed to purchase and use special rights that had been granted to veterans of the Union Army and Navy. For example, Union veterans were allowed to claim 160 acres of federal land within a railroad land grant, while nonveterans could claim only 80 acres, and veterans had the right to buy 160 acres within the railroad land grant for half-price—$1.25 per acre. Veterans frequently sold these rights, sometimes for as little as 40 cents an acre.[20]

Original owners of warrants, scrip, and assignable rights to land very rarely passed those on to the ultimate buyer. Instead, they sold their claims and rights to land brokers, middlemen who gathered government paper from many sources and sold it to land companies and other buyers at a profit.

The railroads were even more important sellers of land on the plains than was the government. Altogether, the railroads received more than 180 million acres of land in the West, about three-quarters of it from the federal government directly, with another quarter coming from the states out of the federal land they had been given. Governments granted these lands to induce the railroads to build on the plains, where existing population and traffic did not justify construction and might not for many years. In addition, railroads purchased lands when they anticipated profits from doing so. They were more clairvoyant than the average settler in their purchases because they knew what route they would be taking and thus which lands were most likely to appreciate in value. The railroads anticipated profiting from their holdings, but the key word in that statement is anticipated. If their land could be sold at a good price and, more importantly, if the land could be made productive, providing substantial traffic for the railroads, they would profit. If the land could not be sold and did not provide shipping business for the railroads, it might actually become a burden.[21]

Selling the land and making it productive were necessary and complementary enterprises for railroads in the West. To sell their holdings they developed divisions or subsidiary companies that marketed the land, aggressively promoting it, luring potential settlers to it, and offering attractive terms to buyers. To make the land productive, they undertook economic development programs, especially those promoting better farming. The railroads introduced new crops and animal breeds, encouraged innovations, sponsored irrigation, supported agricultural science, and cooperated with the other agencies devoted to agricultural development, such as state departments of immigration, land-grant colleges, and agricultural experiment stations. The railroads believed that their interest in making the plains productive and profitable coincided with the interests of the federal government and settlers on the Great Plains. Congress did not always agree and became increasingly critical of the land grants and loans extended to the railroads during and immediately after the Civil War. Farmers and shippers in town also disagreed, and their complaints about high rates and a variety of railroad abuses helped fuel the Farmers Alliance and Populist movements of the late nineteenth century. Still, while the railroads' activities on the plains were clearly self-interested, theirs was an enlightened self-interest. No entity worked harder to develop the region than did the railroads.[22]

The railroads' real estate divisions and subsidiaries sometimes sold land in large parcels to other land companies or syndicates. During the 1880s the Atchison, Topeka, and Santa Fe disposed of much of its Kansas land in twenty- to thirty-section blocks (12,800–19,200 acres), and the Northern Pacific sold large tracts to land companies in North Dakota in the 1870s and again in the 1890s, following its reorganizations from bankruptcy. Land companies that purchased extensive tracts of railroad or federal land usually subdivided it and retailed it to individual farmers, but they sometimes held it for cattle-grazing or crop-raising purposes. The massive bonanza wheat farms of northern Dakota Territory, for example, sometimes totaled thirty thousand or more acres acquired from the Northern Pacific. In Kansas and Nebraska, Englishman William Scully held seventy thousand acres, on which he placed tenants. Large sales provided the railroads with substantial and necessary cash infusions, and it was simpler, easier, and cheaper to deal with one buyer than with one hundred. In general, though, the railroads preferred to sell directly to settlers, either individually or organized in groups. Per acre prices were higher with direct sales, and buyers settled the land quickly and usually densely, providing shipping business to the railroads. Critics of railroad and private land companies frequently accused them of holding land off the market to drive prices higher, but that was rarely the case. The railroads had an incentive to settle the land quickly, and private land companies found that holding land off the market tied up their capital and brought rising property tax liabilities.[23]

The railroads pursued immigrants systematically and aggressively. Their publicity departments churned out hundreds of pamphlets, posters, and newspaper articles, in several languages, extolling the virtues and potentials of their service areas. The language of these publications was florid, and their portrayals and promises were excessive, leading some contemporary observers to damn them as misleading at best and fraudulent at worst. Without question, they were boosting their areas, and they usually put the most favorable interpretation possible on reality. But the railroads, like state immigration bureaus and local commercial clubs—forerunners of today's chambers of commerce—sincerely believed that greatness lay in the future for their regions. Perhaps they deceived those they induced to settle, but they were also deceiving themselves.

The railroads employed agents who worked to recruit settlers. By 1875 the Atchison, Topeka, and Santa Fe maintained sixty immigration agencies, mostly in the East and the Midwest, but promotional activities did not stop at the Atlantic Ocean. In 1883 the Northern Pacific had 831 agents in the British Isles and another 124 in Germany, Switzerland, the Netherlands, and Scandinavia. In most cases, agents and subagents supervised by them received a commission for every immigrant they steered to the railroad. The agents worked for the railroads but provided valuable services to immigrants, acquainting them with the region to which they were moving and helping them arrange transportation, often through steamship companies with which the railroads had arrangements. This process sounds generally positive and benign, but it often reflected the biases of the age. Northern and western Europeans, often defined as "good farmers," were preferred over eastern and southern Europeans. The Atchison, Topeka, and Santa Fe aggressively courted German Mennonites emigrating from Russia, while actively discouraging African Americans interested in relocating to Kansas. Whether this reflected the railroad's bias or its perception of the biases of white Kansans is hardly relevant. The state that stood for free soil in its early years had apparently lost whatever commitment it had to racial equality by the 1870s.[24]

Immigrants from Europe were met at immigration depots such as New York's Castle Garden, and later Ellis Island, by railroad representatives wary of competing railroad agents who might hijack the new settlers. Immigrants, as well as Americans migrating to the plains, purchased "homeseekers' tickets" which allowed them to disembark at several locations to examine available lands. Men accompanied by wives and children could house their families in railroad dormitories while they inspected potential farmlands. The Burlington Railroad maintained "emigrant houses" for families in Burlington, Iowa, and Lincoln, Nebraska, and the Northern Pacific put families up in "reception houses" in Duluth, Brainerd, and Glyndon, Minnesota. While women and children waited, men visited available lands, guided by company personnel who functioned in the way land scouts did in areas open to homesteading, and discussed prices and credit terms with company agents. If they purchased railroad land, the price of their homeseekers' ticket was usually deducted from the cost. In addition, the railroads offered reduced fares and sometimes even free passage for families moving west and for settlers' belongings. Families moving to the plains commonly took an "immigrant car," basically a box car with a coal stove, in which the family, its tools and household goods, and even some of its animals made the trip.[25]

Settlers who bought railroad lands appreciated proximity to transportation and convenience in locating land and moving to it, but the terms of purchase also attracted them. Railroad land prices varied, depending on the quality, location, and level of demand for the land and the strength of the overall economy, but settlers usually found prices reasonable and purchase terms fair. During the 1870s and 1880s, most railroad land sold for four to ten dollars per acre, though the railroads offered discounts of up to 25 percent to those paying cash. Buyers seldom paid cash because they did not have a sufficient amount of it or because they did not want it tied up in land. Borrowers also enjoyed a tax advantage because they were not liable for property taxes until they paid off their loans and received patents from the railroad. Borrowing made a lot of sense under the terms the railroads offered.

Fred and Katharine Kruse bought 240 acres of Union Pacific land near Grand Island, Nebraska, in 1878 for three dollars an acre and were given five years to pay. The Kruses did not put money down, though the railroads usually required 10 to 20 percent. Terms of loans varied a great deal. The Union Pacific usually charged interest of 6 to 7 percent. The Burlington charged 6 percent and gave borrowers ten years to pay—an uncommonly long-term loan in the late nineteenth century. The Northern Pacific leant money on the seven-seven plan—7 percent interest for seven years. In common with other mortgage lenders, the railroads collected only interest over the life of the loan, with the principal due at maturity. So the Kruses paid interest on their loan for five years, at which time they were required to make a balloon payment of $720, covering the full amount of the principal. Because many borrowers could not make that large final payment, they usually refinanced with another lender and used their new loan to pay off the railroad contract. The railroads hoped to profit from their land sales and financing, but they operated in a competitive land market and had a strong commitment to their borrowers' success. All lenders want their borrowers to succeed, but when a settler on railroad land failed the road was doubly damaged: it not only lost money on the loan but also lost a customer who shipped on its rails. Consequently, the railroads tended to be lenient with borrowers and were reluctant to foreclose. That same commitment to nurturing the shipping business also led them to develop liberal terms for squatters on their lands.[26]

Around 1880 Jan Barzyński negotiated an agreement with the Burlington Railroad. The railroad designated him its agent for Polish colonization, and Barzyński committed himself to settling four hundred Polish families on Burlington land in the Nebraska counties of Howard, Greeley, Sherman, and Valley. The railroad promised to reserve three townships for the colonists, provide free passage for them, and donate land for a Polish church and cemetery. The story of Barzyński and the Burlington Railroad reveals that, in addition to the locational and economic advantages of settlement on railroad lands, there were significant social and cultural advantages.[27]

Colonization was a popular method of settlement throughout the plains but was easiest on railroad lands, where colonists could be guaranteed entire townships to themselves. Many colonists were Yankees—established Americans such as utopians intending to live in accordance with a particular set of values, Civil War veterans, or just neighbors hoping that "transplanted neighborhoods" would allow them "happiness and contentment in their surroundings." But many others were immigrants, attracted to group settlement because of the comfort and security it would provide and because it would allow them to more easily maintain the language, institutions, and customs of their home countries.[28]

Colonization was sometimes organized by land companies, sometimes by agents working on commission, such as Barzyński, and sometimes by community leaders such as priests and ministers. The railroads encouraged colonization because of the numerous benefits it offered them. Sales to colonies allowed the railroads to move large parcels of land while assuring themselves that the area would be densely settled. The railroads also recognized that settlers with the support network that came from living among friends and neighbors were less likely than the unattached to become discouraged and leave. This was especially true of immigrants, strangers in a strange land who could not return to their homes as readily as could people with families in the United States. Consequently, the railroads made a major effort to attract and hold colonists. Among other inducements they discounted land, donated land for community purposes, and provided free passage or reduced fares. In 1874, for example, the Atchison, Topeka, and Santa Fe sold one hundred thousand acres of Kansas land to Mennonite colonists. As part of the agreement, the railroad transported the colonists and their household goods for free, "provided free transportation for building materials for one year, and guaranteed aid in the event of any disaster." In Dakota Territory, the Northern Pacific offered German Mennonites from Russia land for the rock-bottom price of three dollars per acre. It also promised them reduced freight rates and passenger fares. Mennonites had the reputation of being excellent farmers, but it was generally true that organized colonists could anticipate attractive terms from the railroads.[29]

Colonization was a win-win situation, as positive for settlers as it was for the railroads. Living among friends and neighbors, colonists enjoyed a sense of security and familiarity that was often difficult to achieve on the strange and sometimes forbidding Great Plains. And they had a ready-made community, composed of people who spoke the same language, had the same aspirations, and were bound together by familiar institutions. The contrast between the colonies, with their population stability and established institutions, and many homesteading areas, with their rapid population turnover, was telling.

Settlers on the plains acquired their holdings in a variety of ways, each of which had its advantages and disadvantages. But all of them shared the experience of farm making, of turning a grassland, however acquired, into a productive farm. For all plains settlers, that endeavor consumed their first decade and sometimes their first generation on the land.

2 How They Built Farms

IN 1894, when Rachel Kahn saw her new home for the first time, she was stunned. The Calof family, into which she was marrying, lived twenty-five miles by wagon north of Devils Lake, North Dakota. They held three claims and lived in three twelve-by-fourteen-foot claim shacks, one inhabited by the parents of her future husband, Abraham, one inhabited by two of his brothers, Charley and Moses, along with Charley's wife and four children, and the third set aside for Abraham and Rachel. Living conditions for her future family were primitive, at best. Her in-laws dressed in rags, had little oil for lamps or fuel for stoves other than animal waste, and had dirt floors and few furnishings in their shacks.

In her first winter on the claim, the Calofs didn't have enough money to heat and light three shacks. Consequently, Abraham and Rachel were joined by his parents and brother Moses, along with two dozen chickens and a calf brought inside for protection from the elements and, perhaps, to provide a little more warmth for the humans. This process was repeated the next year, though now there were six Calofs in the shack, because Rachel and Abraham had welcomed their first child into the world. Another child was born the next summer, meaning that seven Calofs shared living space in the winter of 1896–97. No wonder Rachel noted in her memoir the irony that what she most missed on the wide-open Great Plains was privacy.

Summers were better, but hardly idyllic. Breaking the land was a slow process, and Abraham was frequently away, working for neighbors. In the summer of 1896, gophers ate most of what little wheat they produced, and in 1900 a hailstorm wiped out their crop, damaged their shack, and killed their horses. Another summer a tornado narrowly missed their house, but lightning struck their chicken coop, killing all of the poultry.[1]

The trials of the Calofs, who eventually established a profitable farm, were extreme, but they differed in degree rather than in kind from those suffered by all plains settlers. All settlers faced the challenge of making homes and farms in a strange and sometimes forbidding place, usually with limited resources. This chapter is about how they did it.

Homemaking

One of the great attractions for plains settlers was the opportunity to build a farm from scratch, "to start a farm right once and not take what somebody else has begun," as one farmer put it. But the prospect of creating something out of nothing was as daunting as it was exhilarating. After acquiring their land or a claim to the land, settlers had to move rapidly to build homes to shelter their families and dig wells to draw their water. They had to provide fuel and fencing. They had to break land and put in crops. They had to acquire livestock and implements, if they had none, and build enclosures and shelters for their animals and tools. There was much to be done quickly, but it took years, and sometimes decades, to turn a farm on the Great Plains into a comfortable home and profitable business. As Seth Humphrey wisely noted, "One hundred sixty acres of tough prairie sod do not constitute a farm; they are the raw materials out of which a farm can be made with proper equipment and years of hard labor."[2]

Shelter was the first priority. Settlers who traveled by wagon could live under canvas for a while, but unless they brought tents, those coming by rail needed some sort of housing immediately. As North Dakota pioneer Nina Farley Wishek remembered, the emigrant cars in which many settlers traveled with their possessions were emptied as soon as they reached their destination. "Consequently, there were great piles of boxes, baskets, bundles and household goods of all description scattered along the track where they had been unceremoniously dumped." Getting this baggage covered could not be delayed.[3]

The early settlers in most areas located on river bottoms whenever possible to take advantage of the wood and water available there. Settlers near rivers commonly built log cabins, standard pioneer dwellings in America since the seventeenth century, when they were introduced by Finns in New Sweden (present-day Delaware). Log cabins could be built with the tongue-and-groove method, which conserved expensive nails because logs were notched on the ends to fit snugly with logs laid perpendicular to them. Log cabins were frequently insulated with mud and banked with sod. Milled lumber was usually used for the roof and the floor, though dirt floors and sod roofs were common. Log cabins were relatively easy to erect with available tools and were sturdy and long lasting if built correctly. The main problem with them was that they required logs, which were not available to most plains settlers.

What was abundantly available to settlers was earth, which they frequently used in constructing homes. In hilly regions settlers sometimes lived in dugouts, glorified caves at least partially built by excavating hillsides. Settlers could construct dugouts relatively quickly and cheaply, and they tended to be cool in the summer and warm in the winter. But they were dirty all of the time, wet when it rained, dark, and ill-ventilated. Dugouts were also dangerous; horses or cattle or even inattentive travelers sometimes broke through the roof, and the stovepipe elevated only slightly above ground level could easily be plugged with snow and ice, threatening asphyxiation of those inside. Residents of dugouts usually lined the walls with fieldstones or boards, and they sometimes plastered with mud, but because dugouts were made of earth their walls and roofs were attractive habitats for insects, snakes, and rodents. Dugouts were also hard to locate. "Since its roof was level with the surface of the ground . . . a single sod chimney or narrow stovepipe jutting out of the ground was the only visible sign of a house underneath." This feature made them virtually impossible to find at night. Small wonder that families moved out of them as soon as circumstances permitted. After several years in a ten-by-twelve-foot Kansas dugout with a dirt floor and no windows, Hattie Lee's family of six finally could afford a frame house. They built it on top of their dugout, which they transformed into a root cellar.[4]

Sod houses were more popular and more comfortable than dugouts in most parts of the plains. As the name suggests, sod houses were earthen dwellings built with sod cut from the prairie. The sod house builder first identified a level space, stripped off the prairie grass, and tamped down the earth to provide a hard and uniform space. Then the builder cut sod for the

A family outside their sod house in Custer County, Nebraska, ca. 1888. Nebraska State Historical Society (NebHS nbhips 10631). Used by permission.

walls. The walls of sod houses commonly were three to four feet wide at the base, tapering to twelve to eighteen inches at the roofline. Settlers usually fashioned roofs out of sod laid on a framework of logs or milled lumber. Most sod houses had dirt floors, and settlers commonly plastered the inner walls with mud, often mixed with straw, lime, or sand, to minimize insect infestation and crumbling. While roofs and window and door frames had to be bought or fashioned out of purchased materials, unless they were scavenged from abandoned farms, building a sod house required no appreciable cash outlay by a settler doing his or her own labor. Even that was minimized when neighbors gathered for "sod bees." Their cheapness made sod houses attractive to settlers, as did their relative warmth in winter, coolness in summer, and resistance to fire all the time.[5]

After the settler generation passed from the scene, sod houses came to be regarded with a sort of romantic nostalgia. Those who lived in them did not find them to be particularly romantic. There was, first, the fairly obvious but noteworthy problem of maintaining cleanliness in a house made of dirt, a reality that "did not permit or encourage the housekeeper to be too particular." Grace Fairchild, a homesteader who lived in western South Dakota, refused to live in a soddy because she "didn't want to fight dirt all [her] life, having it drop into the food on the table." Soddy residents could address that problem by stretching cheesecloth along the ceiling, but dirt also rose from the floor below, which turned into dust in dry weather. Some settlers spread straw on the floor to keep the dust down, but that tended to introduce fleas and bedbugs, which brought such maladies as "prairie itch" to their human hosts. Rats, mice, and numerous varieties of insects came through the walls. I. E. M. Smith, one of two female homesteaders holding down claims in Montana, complained of "a species of ants that made a specialty of crawling over us at night." Snakes were also a chronic problem. Smith complained of "snakes that wriggled and snakes that squirmed; snakes that suddenly darted out of the little round holes in the walls and snakes that suddenly darted in again; snakes that glided off a shelf over the bed." A house that was not secure from vermin could hardly be watertight, and residents of soddies suffered when heavy rains turned their floors to mud or collapsed leaky roofs. Smith noted that when it rained heavily the only dry spot in the room was under the table, and that was where they kept the flour. Durability was also a problem. Even a well-built soddy lasted only ten or twelve years. All things considered, it is hard to imagine many residents marking the passing of their sod house with much regret.[6]

Despite their drawbacks, dugouts and sod houses were creative adaptations to the plains environment that provided shelter for many individuals and families in the crucial early years of farm making. But they represented a compromise with genteel expectations and middle-class living standards that many plains settlers were simply unwilling to make. Roderick Cameron's mother flatly refused to accompany the family to Kansas if she had to live in a dugout or a sod dwelling, compelling his father to build a frame house. Because their farmstead was sixty miles from the nearest town, he was required to make eight trips for lumber, each of which took seven days.[7]

For those settlers lacking the time or the money Roderick Cameron's father had, the most popular alternatives to sod houses on the plains were tarpaper shacks. These were square or rectangular structures consisting of milled lumber laid horizontally and nailed to vertical studs. The outside of the shack was covered with tarpaper, which was held in place by wooden slats nailed to the wall. Tarpaper shacks had flat wooden roofs and usually wooden floors, and they were sometimes raised a few inches off the ground to help keep them dry inside. Materials and labor for construction could be purchased for thirty-five to sixty dollars, and a tarpaper shack usually outlasted a soddie. In addition, a shack could be moved relatively easily, which allowed it to be sold separately from the land. But shacks could be dangerously hot in the summer, when the black tarpaper attracted heat. In winter, as the Calof family learned, their lack of insulation made them extremely cold, leading settlers to bank the outside walls with sod or manure. Shacks were dark and drab, with few if any windows, though residents usually tried to brighten the interiors with paper or whitewash. More difficult to remedy was the smell of tar, which permeated bedding and clothing and even altered the taste of food. In any season they were so flimsy that they could blow about in the stiff prairie winds, compelling settlers to anchor them to the ground.[8]

Log houses, dugouts, sod houses, and tarpaper shacks were the most common domiciles in the early phase of plains settlement. But plains settlers were flexible and eclectic in their tastes, and they often married two or more housing forms. Hence, it was not uncommon to see a dugout that had been extended at one end by a tarpaper shack (likely as not appropriated from an abandoned property nearby), a log house with one or two sod walls, or a shack with a sod addition. As author Hamlin Garland noted, in a sentiment that would gain Rachel Calof's ready assent, in the common plains dwelling "there is little chance to escape close and tainting contact . . . In the midst of oceans of land, floods of sunshine, and gulfs of verdure, the farmer lives in two or three small rooms." Plains settlers leaped at the chance to add a few more square feet, architectural consistency be damned.[9]

Most settlers immediately followed the construction of a dwelling by building an outhouse for human waste. Outhouses were not universal in settled agricultural regions east of the Great Plains. In areas with mild climates, it was common for farm families to relieve themselves in woods, bushes, or canebreaks. But the openness of the plains and the harsh winter climate dictated building an enclosure over a hole in the earth. Because the outhouse was not anchored to the ground by a foundation, it could be moved to another location when the hole was filled. Outhouses were not used all of the time—settlers used chamber pots at night and when the weather was extremely cold—and much of the time they were unpleasantly hot, odiferous, and fly- and insect-infested. Moreover, settlers had to be careful about their placement because human waste could foul wells, spreading typhoid and other diseases. Outhouses were unpleasant and potentially threatening to health, but plains families kept them for a long time. Most did not get indoor plumbing until well into the twentieth century, when rising prosperity allowed them to modernize their homes.

Outhouses might be around for a long time, but few settlers viewed their initial dwellings as permanent solutions to the problem of providing shelter. They were temporary expedients that would house the family, allowing energy and capital to be devoted to the primary task of developing a profitable farm. When the latter was achieved, provision of a more suitable dwelling was high on the priority list. The construction of such dwellings was one of the key signs that a pioneer neighborhood was maturing.

Most plains settlers, with the exception of those few fortunate enough to settle on the river bottoms, faced the challenge of providing themselves with fuel and water. They devoted their attention to addressing that problem even as they erected their primitive dwellings. By the time plains settlement began in earnest, most farmers in the North cooked their food and heated their homes with cast-iron stoves, which radiated heat fairly efficiently and could burn coal or wood. Heating a claim shack during severe plains winters, however, was a challenge for any stove. As one settler in northern Dakota Territory noted ruefully in 1885, "We suffered with cold all day. I sat with my feet in the oven most of the day and bowed over the stove. Although we had a hot fire the storm was so cold that my breath came out like smoke . . . Sad is the life in a Claim Shanty of the far west." At least that settler had a hot fire. Many settlers found timber supplies too far distant from their homes or already owned by others, making firewood a scarce commodity. Coal was very expensive, especially when it had to be shipped or freighted in from far away. Grass was abundant, and farmers frequently twisted it into tight knots for the fire. Grass burned well but did not create a fire as hot or as long-lasting as did wood or coal. Settlers also burned corn cobs and animal waste, such as buffalo chips or cattle manure. Importing a practice from their European home, Germans from Russia burned *mistholz*, cow manure mixed with straw and then dried and cut into bricks. Rachel Calof was shocked that a respectable Jewish family would burn animal waste to heat their shacks and cook their food, but she quickly realized that it was the best available alternative.[10]

Heating and cooking with buffalo chips, cow chips, or *mistholz* entailed complications and drawbacks. Gathering enough chips to keep the stove going was a time-consuming task, usually delegated to children. It was also a dirty and a dangerous job because rattlesnakes frequently sought warmth under cow manure. Catherine Porter and her siblings piled gathered cow chips on their northwest Kansas claim and then brought the wagon to take them home. Unfortunately, they "sometimes found when we went for them . . . that someone had stolen them." Then there was the challenge of burning them. When the stove drew efficiently, there was no problem, but when it did not "an unpleasant odor . . . filled the house." Cow chips burned quickly—it was estimated that it took two washtubs full to get through a cold night—requiring frequent stoking of the fire and removal of ashes. There were also sanitary problems involved with bringing animal waste into the house and handling it while preparing food. Seth Humphrey, whose travels as a mortgage company representative resulted in his sharing many a supper with a settler family, was taken aback witnessing "the alternative handling of the fuel and the biscuit dough." As a consequence, he made it a practice to "stroll . . . around out of doors until the meal was ready." Out of sight, out of mind.[11]

Most plains settlers also struggled with the problem of finding adequate water. The lucky few living on year-round streams had water available, but it was usually not potable and required straining and boiling before it could be consumed by humans. The Camerons lived on a stream called Prairie Dog Creek, but as Roderick remembered, the "water was far from crystal-clear. Our full barrel always contained an abundance of bugs and wrigglers. These mother would carefully strain out by passing the water through a cloth." At least for them obtaining water was relatively convenient. Settlers sometimes traveled a mile or more to ponds, sloughs, streams, or neighbors' wells where they could fill water barrels, which they then dragged home on stone boats, low sleds used in the removal of stones from fields. Some settlers developed a partial solution to the water problem by constructing cisterns to collect rainwater. Rainwater was potable, but rain was too infrequent and unpredictable to meet all of a family's water needs.[12]

Long-time residents on the Great Plains joked that "every man . . . has running water three hundred feet from his door." The problem was that it was three hundred feet straight down, necessitating the digging of a well. Well digging was an expensive proposition, especially on the western plains, where groundwater tended to be farther under the surface than was the case to the east. Settlers usually engaged professional well drillers to determine the best site for a well, though some used the services of water witchers, or dowsers. Water witching was a folk practice involving use of a Y-shaped stick to find water close to the surface. A stick that twitched or bent toward the earth was an indication of water nearby. Professional drillers also sought water close to the surface but not so close as to be likely to run dry. At the turn of the century, drillers charged about a dollar a foot, and on the western plains wells could be several hundred feet deep. The expense of a well did not end when the drilling was completed. The sides had to be lined to prevent cave-ins, and a method for bringing water to the surface had to be devised. On the eastern and central plains, where water tables tended to be higher, settlers often dug wells by hand, lining the walls with brick or fieldstone. These were often "open wells," in which water was brought to the surface by the use of a bucket lowered with a rope. Nebraska settler and later Populist leader Luna Kellie drew water from an eighty-six-foot-deep open well with "a long zinc bucket . . . and rope wound around a drum, but it was very hard work for me." In addition to being hard work for the drawers of water, of whom many were women such as Kellie, open wells presented a safety hazard for children and could easily be contaminated by dirt, manure, and animals.[13]

Covered wells were safer and cleaner than open wells, but bringing the water to the surface remained an arduous task. After two years on their claim, Catherine Porter's family was able to afford to have a well drilled; they struck water at 119½ feet. It was a covered well, with a narrow shaft, and the water was collected through use of "a long, narrow galvanized-iron bucket, four inches in diameter, which had a valve in the bottom through which the water came." The bucket was then raised "by a pulley and rope or by a large windlass turned by hand." Raising enough water for family use involved a lot of hard physical labor. Like many plains residents, the Porters also watered their stock from the well. "We would fill two or more tubs and all of our buckets, then bring the cows to the well, and draw water as fast as we could in an attempt to keep up with their thirst . . . By the time we had accomplished this little chore we would . . . be almost exhausted." More affluent families put hand pumps on wells or purchased windmills, but, while these devices saved labor, they were hardly foolproof. As one man remembered of his boyhood on the plains, "When there was not enough movement of air to turn the windmill wheel, cattle would complain noisily, asking for water that was in short supply. Winter was no better. When the pump was frozen there was still a shortage of water."[14]

The time, expense, and labor involved in securing water made plains settlers very conservative in using it. The baby's bath water could be given to animals or used to water the garden. One settler "used the dirty wash water to scrub the floor and carefully poured any remaining water around the family's only tree." Settlers took baths infrequently, especially in the winter, and several members of the family usually used the same bath water. Women washed up with a cloth and a basin, but they sometimes waited "many years before taking a bath in a 'real tub.'"[15]

Living in a sod house or a tarpaper shack, scrounging for fuel, and searching for dependable water were just a few of the adjustments plains settlers had to make. The environment of the Great Plains, with its difficult climate and wide-open spaces, was daunting—so different in degree from what settlers had known farther east or in Europe as to almost become a difference in kind. Every season presented perils. Winter brought blizzards of snow and wind in which whiteout conditions disoriented those caught out of doors and prevented people from seeing more than a few inches in front of their faces. Blizzards could be deadly at a time when dependable weather forecasting did not exist and travelers might be caught in the open miles from home. In January 1888 the "Children's Blizzard" killed an estimated five hundred people, mostly schoolchildren attempting to walk home, in Dakota Territory and Nebraska. As Seth Humphrey noted wryly, "After the first Dakota winter, the prospect of four more like it on a wind-swept prairie sent many a weak-kneed homesteader 'back to his wife's folks.'" Those who stayed learned to take prudent steps, such as locating the door to the house on the east side, where drifts were less prevalent, and running a rope between the house and the barn to avoid getting lost in whiteouts.[16]

Early spring and fall brought prairie fires, which swept across the grasslands rapidly and unpredictably. Prudent farmers cleared fire breaks to protect their buildings, but crops were frequently lost, and voracious flames sometimes jumped breaks and consumed homes and outbuildings. In the early years of plains settlement, prairie fires were sometimes set by Indians burning the grassland to stimulate fresh growth and create a more attractive environment for game or by settlers burning tall grass to make sod breaking easier. Later, lightning, sparks from locomotives, and careless travelers were the main causes of fires. Children were especially entranced by them. A Kansas teenager wrote in her diary in April 1871 that "in the evening there was a splendid prairie fire northeast of us. There is a prairie fire in sight every night almost and sometimes 5 or 6." Coming to North Dakota in the spring of 1886, Nina Farley Wishek and her sister were fascinated to see fires on the

Wahpeton, North Dakota, farmhouse after a blizzard in 1893. State Historical Society of North Dakota (00006-02-2). Used by permission.

horizon every night. But when she narrowly escaped death from one, her fascination diminished, and she noted that "a big fire seems to create wind as it advances, causing a veritable inferno or whirlwind of fire."[17]

Spring and summer brought tornadoes, cloudbursts, and hailstorms like the one that devastated the Calof farm. The growing season was also accompanied by hordes of gophers and rabbits and by plagues of insects, such as mosquitoes—called "Platte River birds" and other colorful, if less printable, names—and chinch bugs, which ate the roots of wheat and other grasses. But the most frightful adversary of plains settlers was the Rocky Mountain locust, more popularly known as the grasshopper.[18]

Grasshoppers were particularly virulent on the eastern Great Plains in the 1870s, though major infestations bedeviled plains locales well into the 1930s. Kansas was especially hard hit in 1874. One immigrant woman there wrote that "here came millions, trillions of grasshoppers in great clouds . . . eating up everything . . . the leaves on the trees, peaches, grapes, cucumbers, onions, cabbage, everything . . . Only the peach stones still clung on the trees, showing what had once been there." Another Kansan noted that, "when they came down, they struck the ground so hard it sounded like hail." One Bohemian settler in eastern Nebraska remembered that, in 1875, "we heard a sound and it grew dark and we thought that a storm was coming. The sun was hidden. We thought that it was the end of the world. Then they began to come down. In one hour they had eaten everything . . . They were so thick on the ground that when we took a step they were over our ankles and our feet made holes, like footprints in the snow . . . They would eat the paint off a house and chew up lace curtains." Nebraskan Luna Kellie recalled that "they lit on my head my dress my hands and no place to put my feet . . . And the noise. Who could believe a grasshopper feeding made a noise but the whole array of them made a noise as of a band of hogs chanking." Grasshoppers were a gift that kept on giving long after they moved on. "Everything reeked with the taste and odor of the insects. The water in the ponds, streams and open wells turned brown with their excrement, and became totally unfit for drinking . . . Bloated from consuming the insects, the barnyard chickens, turkeys, and hogs themselves tasted so strongly of grasshoppers that they were completely inedible."[19]

The Great Plains was a difficult environment, but the settlers inadvertently made it worse than it was naturally. Gophers and rabbits thrived in part because of the unrelenting war settlers waged on coyotes, wolves, snakes, foxes, hawks, and other predators. Large grain fields presented an attractive environment for chinch bugs. Russian thistle—or, more popularly, tumbleweeds, an invasive species accidentally introduced by farmers—choked out native grasses and allowed prairie fires to spread more quickly. Even grasshoppers thrived in part because plowed fields provided the perfect bed in which they laid their eggs.[20]

Plagues of fire and ice and locusts of almost biblical proportions, when joined with the other difficulties of farm making on the plains, dissuaded many settlers from staying. They went back home or moved on to another opportunity. Leaving was easier for some than for others. When one winter in a shack "permanently chilled" Hamlin Garland's "enthusiasm for pioneering," he abandoned his claim, but he was an educated American professional with lots of options. Immigrants who spoke no English and had no skill but farming had fewer alternatives, though they, too, found the plains daunting, as did O. E. Rölvaag's fictional heroine Beret, who concluded that "human life would not endure in this country." But moving elsewhere or moving back was not always a viable option. Some settlers, such as the Calofs, could not even consider going home. When Jews got out of Russia, they did not go back. Rational people do not break into prison.[21]

Women, such as Beret and Rachel Calof, seemed to suffer more from the difficulties of the plains environment than did men, though that proposition may be more apparent than real. Certainly, women were more literary than were men, leaving us reminiscences and memoirs of plains life that frequently stressed its hardships. Moreover, contemporary journalists and authors tended to emphasize the sufferings of women on the plains and to praise their courage and valor. Hamlin Garland, for example, highlighted the distress of women on the plains in much of his work, including both fictional treatments and nonfiction reminiscences focusing on his mother's travails on the farm. The isolated and desolated farm wife was a staple in the reportage of eastern journalists who visited the Great Plains. Those who knew the plains well, such as Seth Humphrey, also focused on the unhappiness of farm women; he suggested that much of the abandonment of farms in the 1880s "was due to the women."[22]

The standard portrayal by contemporary novelists, journalists, and other observers of plains women as unhappy victims became a stereotype that obscured as much as it revealed. Loneliness, depression, and even despair could be found among the women in soddies and claim shacks, but so could optimism, enthusiasm, and pride in creating homes and farms. Especially among single female homesteaders, like I. E. M. Smith or Mina Westbye, who made up one-sixth or more of the claimants on the western plains after 1900, there was a sense of adventure and a thirst for independence that made them anything but downtrodden and depressed.

There were differences, however, between single and married women on plains farms. The former were moving beyond rigid gender expectations and stereotypes, while the latter continued to be circumscribed by them. There were three cultural expectations in particular that applied to most married female settlers on the plains. First, though decision-making processes varied from family to family, women were expected to accept their husbands' decisions and to move their families, even when that meant leaving neighbors and kin and relocating to an unfamiliar environment. They did not always comply, or they imposed conditions, as did Roderick Cameron's mother, but the expectation was that women would comply. Second, they were expected to make homes, bear and raise children, cook, clean, construct and care for clothing, and contribute to the economic success of the farm, regardless of the paucity of necessities, such as water and wood, and of conveniences that would facilitate the accomplishment of their work. Third, while men could go to town to market crops, get tools mended, or purchase supplies, women were expected to remain at home caring for young children. Women were thus more likely than men to be isolated on plains farms, especially in the early days of settlement.

It was this isolation that seemed to preoccupy them most. Neighbors were often a half-mile or more distant, a reality captured by a Kansas farm woman who wrote that "I never saw a light from a home at night all the time we lived on the farm." Distances made visiting neighbors or going to town difficult at any time and virtually impossible in the winter. Mary Dodge Woodward lived only eight miles from Fargo, the most significant town in northern Dakota Territory, but was not able to go there in the winter. Meanwhile, "if the men get out of beer, they go in any weather." Likewise, in the Nebraska Sandhills of Mari Sandoz's youth, "The men could get away from their isolation. They could go to the warm, friendly saloons . . . But not their women. They had only the wind and the cold and the problems of clothing, shelter, food, and fuel." Not only was isolation socially stultifying and psychologically damaging, it could also be life-threatening. That reality came home to Rachel Calof when, during a difficult pregnancy and again after a child was badly burned, she reflected that it took a day and a half to get to a doctor and another day and a half to get him to their homestead.[23]

The special challenges for married women presented by life on the plains extended to their customary domestic duties. Making and keeping a house in a new and formidable environment were as difficult as making and keeping a farm. Maintaining cleanliness, always a problem on farms, was especially difficult in dugouts and sod houses. Cooking with insufficient fuel and food was an ongoing struggle, and when there was enough food women faced the challenge of adding variety to a monotonous, high-starch and high-fat diet. Keeping children safe on a prairie where they could easily get lost or stumble onto rattlesnakes and other hazards was an endless preoccupation. Even washing clothes became a major undertaking; women made their own soap out of lye and animal fat, carried water for long distances, heated it with insufficient fuel, and scrubbed and boiled garments.

Women making homes on the plains might have been in the wilderness, but they had no intention of being *of* the wilderness. In fulfilling their homemaking duties they were "carriers of domestic ideology," transferring what they understood to be civilized standards and behaviors from Europe or the East to the Great Plains. This meant a commitment to attempting to maintain standards of cleanliness, nutrition, dress, behavior, and language. It meant reproducing, as soon as possible, schools, churches, social networks, and the other formal and informal institutions of genteel life. It often meant keeping books, musical instruments, and other "finer things in life" in the home, and it meant applying what was called a "woman's touch" to homes and farms by planting flowers, hanging curtains, decorating, and arranging furniture. Women made their marks in even the most bleak and unpromising surroundings. Rachel Calof, for example, domesticated her claim shack by plastering and whitewashing the walls and by fashioning curtains out of flour sacks. Tokens of gentility often assumed a symbolic importance that transcended the tangible object involved. As Faye Lewis recounted of her mother, a settler's wife in southern South Dakota, "Mother's china was more than a set of dishes to her, more than usefulness, or even beauty. They were . . . a reminder, that there were refinements of living difficult to perpetuate in rugged frontier conditions, perhaps in danger of being forgotten." Lewis's mother, in common with most other plainswomen, had no intention of forgetting.[24]

Farm Making

Creation and maintenance of a home were essential but secondary to the main purpose of most settlers' endeavor—developing a productive and profitable agricultural enterprise. In so many facets of life, timing is everything, but in farming, where nature sets a particularly rigid timetable, the pressures are especially intense. As a student of settlement in western South Dakota noted, "New farms had to race against the clock and the calendar." That was especially true at a farm's inception. Most crops had to be planted in the spring, in just a few weeks' time, which meant that farmers on new land had to be on site and ready to go shortly after the frost went out of the ground. But this was not easy to do. Settlers commonly chose land during the relatively benign months from May through September, which meant that the planting window was usually closed by the time they took possession. As a consequence, they bore the expenses of owning a new farm—filing fees, down payments, the cost of building a dwelling, transportation of goods and family—but enjoyed none of the profits.[25]

In the first year on the land and for years thereafter, the primary task of plains settlers was preparing the land for farming by breaking virgin grassland for crops. Breaking the "hard, compact, and tenacious" prairie land was an arduous, expensive, and time-consuming process because it required cutting through the dense and deep root structure of the prairie grasses.[26] Breaking was a two-step process. Initially the breaker cut the sod to a depth of about three inches with a moldboard plow, the curved sides of which turned the soil on either side of the cut. The grass under the turned sod would rot, and in a few months the broken sod would be "backset." Backsetting involved plowing the broken sod perpendicular to the furrows, one to three inches beneath the depth of the original cut. Backsetting accelerated decomposition of plant matter and helped smooth the seedbed for crops.[27]

Sometimes farmers broke land cooperatively, sharing work with friends and relatives, or in "breaking bees," when a group of neighbors would go from farm to farm. However they attempted to break the soil, settlers quickly discovered that a conventional plow drawn by farm horses was inadequate for the task. Farmers either turned the soil by hand with a shovel or, much more commonly, purchased a heavy "breaking plow" or hired a skilled sodbuster. Even with a breaking plow pulled by heavy oxen, it was difficult to break more than about an acre a day. This made the task time-consuming for farmers and expensive when they hired it done. It was estimated that a professional sodbuster charged two to three dollars per acre to break virgin prairie. Backsetting was less arduous and could be done using lighter plows. Farmers who chose to hire backsetters usually paid one to two dollars per acre. Even when farmers did their own backsetting, the cost of preparing the land frequently exceeded its sale price.[28]

After the land was broken, backset, and harrowed (passed over with an implement with spikes or teeth designed to pulverize clods and level the ground), it was not necessarily smooth enough for all crops. Plains settlers commonly planted corn and flax on virgin prairie. Corn did not require a smooth seedbed and could be planted and harvested by hand. It was not an important cash crop, but it could be fed to animals or consumed by humans "fresh or as grits, mush, fritters, hominy, or corn bread." Flax grew well on rough soil, and its seed—called "linseed"—was in high demand by manufacturers of paint and varnish. The problem with flax was that the crop carried

Breaking sod with a hand plow near Zahl, North Dakota, about 1910. State Historical Society of North Dakota, William E. (Bill) Shemorry Photograph Collection (1-16-27-1). Used by permission.

funguses that remained in the ground, eventually making the soil "flax sick" and dramatically reducing yields.[29]

As plains farmers slowly broke more land, they added more crops to the mix. The Great Plains eventually came to be called the "wheat belt," but, although wheat was the premier cash crop, the region was characterized more by diversification than by monocropping. Diversification was wise because it diminished the high risk of putting all of one's eggs in one basket, a strategy that could be undone by low prices, an insect invasion, or an unanticipated weather event. Successful diversification required some careful planning. Growing several crops that all demanded close attention at the same time could result in significant losses. Plains farmers produced wheat and other small grains, such as barley and rye, for the market. But they also devoted a substantial amount of their acreage to the production of crops for family consumption. In addition to producing garden vegetables that were mostly consumed fresh, such as lettuce and tomatoes, farm families grew corn, potatoes, and root crops (e.g., carrots, onions, turnips, and beets) that could be kept for long periods, beans that could be dried, and cabbage that could be turned into sauerkraut. They used some of their land for growing hay and oats for horses, which succeeded oxen as the draft animals of choice on plains farms because of their greater speed, nimbleness, and versatility. It was estimated that a work horse consumed five or six acres of hay and grain per year, and a farmer needed one horse for every fifty to sixty acres farmed. Virtually all farmers raised other animals as well, including swine, poultry, and sometimes sheep. Cattle, however, were the preferred livestock. Plains farmers especially favored "dual purpose" cattle that could be milked and raised for beef, much to the consternation of animal scientists, who argued that dual-purpose cattle were not particularly good for either purpose. From early spring through late fall, farmers grazed their cattle on the public domain and on their own unbroken acres, confining them in corrals close to barns when winter brought severe weather. Only about 30 percent of the Great Plains was ever broken for crops, from about half in the east to under 20 percent in the west, leaving plenty of natural prairie for grazing animals.[30]

As they expanded their production, farmers discovered that they needed more implements, work horses, barns, machinery sheds, stables, and chicken coops. They had to fence fields and gardens to protect crops from grazing animals. And they found it beneficial to build root cellars and sometimes smokehouses. These were all necessary investments, but they were not always immediately justified by the productivity of the farm.

Farm making was a demanding process that often stretched over many years. Those who undertook it usually had the advantage of an agricultural background, and by the turn of the century many had experience farming on the plains. They had a good sense of which crops would grow and which would not. They knew how to handle farm animals, build fences, construct simple buildings, and repair tools. They knew how to butcher, make sausage, pickle and preserve vegetables, and do all of the other things their farming parents had done. But some settlers had little or no farming experience. The Calofs knew virtually nothing about farming, for the simple reason that historically Jews in Russia had been prohibited from owning farmland. After the turn of the century, the proportion of settlers without agricultural experience appeared to grow, with more than half of the homesteaders in some areas of the western plains lacking a farming background. Some of these people were inspired by a vibrant and active national back-to-the-land movement that portrayed the farm as a sure path to contentment and prosperity. Others were townspeople—teachers, clerks, retailers, and professionals—who saw a homestead as a means of increasing personal wealth rather than as a long-term commitment. The latter usually succeeded, commuting or relinquishing their claims for a profit, but the former usually failed. Farming experience did not guarantee success on the plains, but its absence tilted the odds toward failure.[31]

For serious plains settlers trying to create farms, hard and persistent labor by all members of the family was the rule. Men were generally responsible for field work, while women cared for the home. This meant not simply housekeeping and child-rearing but also hauling water, securing fuel, maintaining gardens, and preserving food. It usually also included keeping poultry flocks, gathering eggs and slaughtering chickens, milking cows, separating cream, and making butter and cheese. Doing poultry and dairy chores—"barnyard" as opposed to house or field work—was generally considered appropriate for women, but it was physically demanding. Handling cattle was not easy, especially when they decided to go somewhere other than where their handlers wanted them to go or do something other than what their handlers wanted them to do. Milking twice a day, in all weather and conditions, involved shifting and uncooperative animals, swinging tails, flies, and chapped fingers and teats. Then there was the demanding but sensitive work of separating cream from milk and turning it into butter or cheese. Poultry production also had its drawbacks because "chickens impose demands that are incessant and the required labor is considerable." The fowl had to be fed and eggs gathered, including those the hens hid. "The labor becomes more burdensome when it involves the malodorous job of cleaning the henhouse . . . delousing the fowls . . . whitewashing and disinfecting coops and nests and roosts" and "medicating for roup, limberneck, and chicken cholera."[32]

Women did most of the house and barnyard work, but they also labored in the fields when necessary. Immigrant women, especially from Germany and Poland, often did field labor as a matter of course, sometimes earning the contempt of Yankees who had embraced a more domestic ideal for women. American-born Nina Farley Wishek viewed field labor by her German Russian neighbors as "an insult to my sex," but she later "came to realize that the girls and women enjoyed the freedom of outdoor life." Whether or not they enjoyed the freedom of outdoor life, they recognized that their labor was necessary for success in farm making, and they often took pride in their contributions to that endeavor.[33]

Farm making was an enterprise in which the whole family was involved. Single men and women could and often did claim homesteads, but it was especially difficult for a single person to create a productive and profitable farm without the help of family members. Children as well as adults worked to make farms successful and were economic assets in the labor-intensive agricultural regime of the late nineteenth century. Children as young as four or five could fetch eggs, pick bugs off of potato plants, and do some light weeding. When they were a few years older, they could draw water, fuel the stove, and pick out some of the smaller rocks heaved up when the fields thawed every spring. Boys and girls in their early teens could serve as full hands, doing the same work their parents were doing. There was no shortage of farm children from the plains who bitterly remembered this work as drudgery. Others took pride in their abilities and their contribution to the family's success. Most recognized that their labor was a necessary part of the family's struggle to carve viable farms out of the plains.

Coping with the Costs of Farm Making

One of the great attractions of the plains was the free or cheap land that was available. Small wonder that settlers were often surprised by how expensive their farms turned out to be. Initially, and usually for some years thereafter, outgo far exceeded income on plains farms. Down payments had to be made, locators had to be paid, and fees had to be surrendered. Dwellings had to be erected, wells had to be dug, and land had to be broken. Implements and horses had to be purchased, outbuildings had to be constructed, and fences had to be strung. None of this stuff was cheap. Fencing forty acres usually ran over three hundred dollars, and a yoke of oxen or a team of horses could easily cost two hundred dollars. Then there were the unanticipated problems, such as crop failures and animal deaths, and emergencies such as fires or illnesses. Farmers with little capital, like the Calofs, could compensate to some degree through hard work, but everyone needed some capital. The Atchison, Topeka, and Santa Fe urged immigrants to Kansas to have a nest egg of eight hundred to one thousand dollars. Another observer estimated that, while a settler might get by with less "through luck, grit, sacrifice, and careful management," he or she should have about one thousand dollars to get properly established. To put that in perspective, one investigator estimated that in 1892 the average working man in Milwaukee made nine hundred dollars *in a year*. Land on the plains might have been free, but living on it hardly was.[34]

Settlers stretched their precious dollars in a variety of ways. If a town or settlement was nearby, women commonly bartered their eggs and butter for store goods or credit. At times this was the major source of income for pioneer farms, without which many more would have failed. Neighbors traded work with one another, and shared implements and horses. Luna and J. T. Kellie rented barn space to passing freighters and sold them feed for their horses. During the winter, men hunted or trapped animals and cut wood or ice for sale if they were fortunate enough to have access to some. Old Jules was especially resourceful, hunting, trapping, locating land, surveying, and castrating cattle and hogs for his neighbors. In return they broke land for him and did his field work.[35]

Barter helped settlers make the most of limited resources, but it was not sufficient to keep a money-losing farm afloat. Like most small farmers today, plainsmen supported their farms by working for wages. Whenever possible, men preferred to work in the neighborhood so that they could care for their families and tend to their farms, though that could be difficult because the local demand for labor was usually highest when a settler's farm needed his fullest attention. Abraham Calof did farm labor for a neighbor for twenty-five dollars per month. He was too far from home to return every day, but his wages paid for flour, coal, and other necessities. J. T. Kelly broke sod and threshed neighbors' grain for wages, while others built shacks for new settlers. James Leslie "walked to Mitchell [South Dakota] every day to operate a mill and livery stable." Frequently jacks-of-all-trades, plains settlers were opportunistic in the search for jobs. In northeastern Colorado, farmers moonlighted as "sugar factory workers, grain elevator assistants, brick factory operatives, barbers, . . . cooks, . . . mail carriers, . . . threshers, road workers, railroad workers, cheese makers, ranch and farm hands, sod breakers, carpenters, well diggers, well pipeline installers, stone masons, cement block makers, fire guard plowmen, cattle skinners, gopher exterminators, sheep shearers, house plasterers, tree claim plowers and planters, and buffalo bone collectors."[36]

Men could not always find work close to home. Kansas settler Henry Martin lived in town during the winter, working at a livery stable and taking other odd jobs. After claiming a homestead in northwestern North Dakota in 1904, Anders Svendsbye spent the winter in the Minnesota woods working as a lumberjack. He followed that pattern for three years while slowly developing his farm. One of Roderick Cameron's neighbors left northwest Kansas for Leadville, Colorado, where he and some friends had a contract to make railroad ties. Leaving home for long periods was not a major problem for single men, who often outnumbered families in newly settled areas, though covetous neighbors could accuse them of abandoning their claims. Married men usually didn't have to worry about residency challenges, but when they worked in distant places their absence increased the responsibilities of their wives. In addition to being responsible for the home and the barnyard, wives were called upon to care for animals and sometimes cultivate and harvest crops.[37]

Single female settlers also usually held off-farm jobs, as did Mina Westbye, who went to Minneapolis every winter to work as a maid and a seamstress. Especially after the turn of the century, when it became more common for women to homestead, the western plains were filled with teachers, librarians, milliners, and others who worked in town while claiming homesteads. Most of these homesteaders did not intend to make farms and were instead holding claims as investments. But single women sincerely interested in farming also took jobs. They enjoyed the advantage of having domestic skills as laundresses, cooks, and seamstresses that were in high demand in towns. After two sets of cousins—Clara and Mary Troska and Helen and Christine Sonnek—claimed land near Bowbells, North Dakota, for example, they quickly secured jobs in town as cooks and servers in the local hotel. It was less common for settler wives to work for wages because they usually had children to care for and heavy farming responsibilities and because a working wife was often taken as a sign of a husband's inability to support his family, but it was not unheard of. Mary Anne Leslie operated a millinery shop in Mitchell, South Dakota, and Mary Dodge Woodard hired the young wife of a settler to keep house, noting that she and her husband "are very poor so that four dollars a week was some inducement."[38]

Children also frequently worked for wages to support the family farm. Older boys were especially valuable as farm hands working for neighbors. Older daughters often worked in hotels, restaurants, and laundries and as maids for affluent neighbors or families in town. Their earnings were meant "to help the father struggle out of debt, or to make it possible for the younger children of the family to go to school." One Kansas teacher, for example, gave $45 of her $125 salary for the year to her father to help him buy horses. The next year she surrendered $60 of the $180 she earned when he needed a new wagon. Sometimes the labor of children was necessary for family survival. After their father died, Catherine Porter's brother went to work for neighbors, usually receiving grain or meat for his pay, while her mother wove carpets for sale in town. Two of Hattie Lee's brothers took work as shepherds to support the family, while she embarked on a series of jobs as a cook, maid, hotel waitress, and cattle herder. The experience left her bitter. "When a child is out in the world . . . no one knows what a hard time she has. No one can tell the hardships I went through; carry water from the well, rub the clothes on a wash board and keep the house work up. And some were so unreasonable," she concluded, "they never thought a young girl ever got tired."[39] These efforts were heroic and necessary, but too often they merely enabled a family to scrape by. Like other Americans, plains farmers seeking success depended on credit.

3 How They Got Credit

In the spring of 1908, North Dakotans Thor and Gjertru Birkel borrowed six hundred dollars to commute their homestead and replace a house that had burned to the ground. This loan turned out to be just the beginning of their involvement in credit markets. Over the course of the next six years, they raised several thousand dollars by mortgaging their homestead at least eight times. Several hundred miles to the south, in Kearny County, Nebraska, J. T. and Luna Kellie waited until they proved up to borrow. At that point they mortgaged their farm for eight hundred dollars because "we had so run behind."[1]

Neither the Birkels nor the Kellies were especially unlucky or impecunious. As Luna Kellie noted, they raised their crops with few horses and rudimentary machinery, she "made all of our household expenses from chickens[,] garden etc," and J. T. worked numerous jobs off the farm. Yet they still could not get by.[2]

The experiences of the Birkels and the Kellies were common on the Great Plains. Even homesteaders receiving free land needed a minimum of five hundred to eight hundred dollars just to get through their first year of farming. Work off the farm, heroic self-sufficiency, and extreme self-denial might allow homesteaders to scratch out subsistence farms after a few years, and if they were fortunate enough to avoid getting sick or having a house burn down or losing the cows in a blizzard, borrowing might not be necessary. But many homesteaders were not able to avoid such unexpected reverses. Moreover, most homesteaders committed to more than proving up and selling out for a profit needed to develop more than a subsistence farm. They sought commercial success, which meant producing a cash crop in sufficient quantities to allow them to improve the lives of their families. In the 1880s in the eastern Dakotas, it was estimated that filing on a homestead, buying horses and implements, building a house and necessary outbuildings, digging a well, breaking 160 acres, fencing it, seeding it to wheat, and harvesting and threshing the grain would cost $3,150, or nearly $20 an acre. To raise the sort of money required for commercial success, the farmer entered into what Kansas pioneer Roderick Cameron called a "'vicious cycle.' To produce more, he had to increase his acreage. To do that profitably, he needed labor-saving implements. To acquire these, he had to have more cash capital, and that meant—a farm mortgage," which had to be paid off with still greater produce and more profits. One way or another, then, most plains farmers borrowed, often several times from different types of lenders.[3]

Short-Term Credit

The American economy has always been fueled by credit, and that was particularly true in the late nineteenth and early twentieth centuries. At every stage of economic endeavor, participants were enmeshed in a web of credit. Retailers were carried by wholesalers who were carried by manufacturers who were carried by bankers. Railroads, which owed their very existence to bondholders, bought rolling stock, rails, ties, and structural steel on credit and in turn extended credit to shippers, farmers, and purchasers of their land. Farmers, at the bottom of one or more credit chains, borrowed from a variety of lenders for a variety of purposes.

The Great Plains region in the late nineteenth and early twentieth centuries was even more credit dependent than was the nation as a whole. In a sense it was similar to a high-tech startup today. It promised impressive profits in the future, but it wasn't profitable yet, and it required large infusions of capital in order to realize its promise. The farmers who formed the base of the economic structure on the plains generally needed two types of credit, short- and long-term. Short-term credit was usually for a year or less but could be extended for up to a two-year period. Farmers regularly sought short-term credit for production goods, such as draft animals, implements, and seed, but they sometimes required loans to feed and clothe their families or cover emergency expenses. Many farmers needed short-term credit every year, taking out loans at planting time and just before harvest and paying them off after harvesting and marketing their crops and animals. But demand for credit rose when low prices or natural calamities reduced their incomes beneath their expenses. That was usually at the very time when local lenders were not in a strong position to extend credit.[4]

Farmers obtained short-term loans locally and, as was the case with all lending, family members often provided the money. Thor Birkel, for example, borrowed money from his brother Kristen for horses, implements, and harness. Even after he embarked on a course of borrowing from financial institutions, he continued to seek assistance from Kristen. Borrowing from family members undoubtedly caused frictions and hard feelings, but it was difficult for relatives to refuse and they were more likely than outsiders to be forbearing with the borrower.[5]

Settlers also borrowed from general stores. Local merchants expected to operate largely on credit. General stores traded necessities for cream, butter, eggs, pelts, and other items. They also carried farmers until harvest if they lacked goods to barter or ready cash to pay their bills. Because this was standard business practice, any merchant attempting to operate solely on a cash basis could not succeed. At the same time, it was risky for local merchants to extend credit to their customers because they were financially obligated to wholesalers and because their customers were sometimes unable to pay. To compensate for the higher risk of credit sales, merchants frequently had two prices for the same item—a "credit price" and a lower "cash price."[6]

Farmers who borrowed beyond their circle of family and friends or who needed more credit than the local general store could provide usually had to take out a mortgage. A mortgage is a legal document in which a lender, or mortgagee, agrees to lend a sum of money to a borrower, or mortgagor, who agrees to repay the loan at a specified rate of interest. The borrower also agrees to provide the lender with security in the form of property, called collateral. In addition to specifying the amount of the loan and the rate of interest and identifying the collateral, a mortgage details such terms as the length of the loan and the frequency and amount of payments the borrower will make. Should the borrower fail to fulfill his or her part of the contract, the lender can take the borrowers' collateral as compensation.

Farmers seeking short-term financing to buy something more substantial than the general store offered—a reaper, for example, or a team of horses—usually had to take out a chattel mortgage from a lender. A chattel mortgage was not secured by land but by movable property, such as livestock, wagons, or farm implements. Lenders were reluctant to contract chattel mortgages because they had their own obligations, demand was usually highest in difficult times, chattel property tended to die or wear out, and movable property was, well, movable. On the other hand, the alternative to extending this sort of financing was often having no business at all. Bankers, implement dealers, and others who regularly contracted chattel mortgages compensated for their risks by charging high interest, sometimes up to 36 percent per year. Some commentators considered chattel mortgages to be especially unjust to borrowers. Seth Humphrey observed that "local bankers . . . charged three per cent a month and were as merciless as crocodiles in cases of failure to pay." He "often wondered why some of these chattel men were not killed" by "some desperately hard-up settler." For their parts, chattel lenders believed that the scarcity of capital in newly settled areas and the high risk they assumed justified the interest rates they charged.[7]

Long-Term Credit

Chattel mortgages were insufficient to meet the credit needs of plains settlers. For large expenditures such as purchasing land or making major improvements to homes or farming operations, mortgages secured by farmland were necessary. Many of those mortgaging farmland were buyers of the land, but it was common for those who already owned land to mortgage it. Homesteaders frequently mortgaged their land. This was easiest to do when they had received title, but many homesteaders mortgaged their land as soon as they proved up. Nor was it unheard of for homesteaders to secure mortgages earlier in the process, even though by law homestead claims could not be encumbered. I. E. M. Smith observed that, in her Montana neighborhood, "everybody . . . mortgaged his land in order to obtain money to prove up." Lenders offering credit to borrowers who did not yet have title to the land on which they were borrowing were taking a greater risk than those lending to borrowers who owned the land outright. They compensated by charging higher interest rates and offering less liberal terms.[8]

Purchasers of land could choose among several sources of credit. The railroads financed buyers of their lands at terms that were generally reasonable. Though rates varied, the railroads commonly charged 6 or 7 percent interest in the early phases of settlement, when rates from other lenders sometimes approached 10 percent. The difference was attributable mainly to the facts that the railroads themselves could usually borrow more cheaply than other lenders and that liberal terms enhanced the likelihood of borrowers' success. While the railroads hoped to profit from the sale and financing of land, their main interest was in filling the countryside with prosperous farmers who would use their services. Their focus on the future of their service areas also encouraged them to be forbearing with delinquent borrowers and to extend loan repayment periods when their own financial circumstances allowed it.[9]

Land companies provided financing to settlers buying land from them, and after the first few years of settlement most communities had local investors who provided loans to farm buyers. In southeastern Nebraska a high proportion of lenders lived in the same county as borrowers and "individuals were the commonest type of lenders of . . . mortgage credit to farmers." Some of these lenders were farmers selling their farms by taking a mortgage from the buyer. This form of lending was especially prominent in areas that had been settled for a generation and in which early settlers were retiring. Other lenders were affluent local people who invested in real estate. Local investors were more likely than outsiders to have a good understanding of the value and productive potential of the land, as well as the agricultural abilities and character of the borrower. National credit agencies such as Dun and Bradstreet and Standard and Poor's provided information regarding the creditworthiness of borrowers, but there was no substitute for local knowledge. The way a man's house was maintained, the condition of his fields, his personal habits, even the way he dressed his children told a potential lender more about his character than a credit report ever could. Borrowers also frequently preferred to borrow locally from someone they knew, believing that a local person would deal with them fairly and sympathize if they had difficulty making payments. Moneylender Herbert Martin's business flourished because "our fellow Kansans, by and large, would rather deal with a man than a bank."[10]

Relatives also provided long-term credit. For the borrower, the family offered all the advantages of the neighborhood lender with the added benefit of ties of blood. In addition to helping relatives get started or advance their enterprises, families used mortgages to pass property from one generation to the next. Aging parents frequently held mortgages on land transferred to children. The mortgage protected other heirs to the estate, and regular payments on it provided parents with a steady income.[11]

Local lenders and family members were insufficient to satisfy the voracious demand for capital on the rapidly developing Great Plains, especially during the early settlement boom. That demand was met by other institutions, the most significant of which in the 1870s and 1880s were mortgage companies. Mortgage companies functioned as middlemen between lenders and borrowers, moving money from where it was being earned and saved—the Northeast and Europe—to the plains—where it was needed and could be invested relatively safely and profitably. Mortgage companies were sometimes family partnerships, such as that formed by the Davenport brothers of Bath, New York, and sometimes private companies. By the 1880s most were incorporating to increase their operating capital, expand their businesses, and more effectively meet their clients' needs. Virtually all of these companies were headquartered in New England, the Mid-Atlantic states, or Europe.[12]

Mortgage companies initially operated as brokers, arranging loans from affluent individuals to western farmers. The company would identify a potential borrower, appraise the property to be mortgaged, perform due diligence regarding his or her abilities and creditworthiness, write up and execute the loan agreement, assure that the borrower paid taxes and otherwise behaved responsibly, and collect loan payments. In return, the company received a commission from the borrower and a loan-servicing fee from the lender. Lenders liked this system because they could earn a higher return than they could earn on mortgage loans in the East. Moreover, investment in farm mortgages seemed relatively safe because brokers such as J. B. Watkins had established reputations for probity and fiduciary responsibility, and the companies usually agreed to purchase delinquent mortgages, diminishing lenders' risks.[13]

In the 1880s some mortgage companies expanded their product line by adopting the debenture system. Debentures were interest-paying bonds secured not by a single mortgage but by a large number of mortgages grouped together into trust accounts of $100,000 each. The debenture system was attractive to mortgage companies because it was less expensive to operate than the brokerage system and was thus more profitable. Smaller investors, who could not afford to carry an entire farm mortgage, could invest in debentures, and the fact that the debenture was secured by a large number of mortgages made it seem safer. The problem with the debenture system was that the mortgage companies and their local agents (usually lawyers, realtors, and bankers), whose incomes depended on the commissions they generated, were tempted by it to make riskier loans than they could make under the brokerage system. A questionable mortgage that could not find an individual buyer could be hidden among the dozens of higher-quality loans in a trust account. Brokered loans continued to dominate the farm mortgage market on the plains, but by the late 1880s a little under 20 percent of loans were being made under the debenture system.[14]

Terms and Conditions of Farm Mortgages

Farmers on the plains discovered that borrowing money from a nonfamily source was a complicated undertaking. In the lending and borrowing process, virtually everything was negotiable, but the major points of contention between lenders and borrowers were usually the length, amount, and interest and commission rates of the loan. Two considerations played a major role: (1) the supply of and demand for money and (2) the risk to the lender and the borrower.

A property owner with a mortgage today expects to have fifteen to thirty years to discharge his or her obligation, but in the late nineteenth and early twentieth centuries mortgages matured more rapidly, usually in two to seven years. Borrowers paid interest on the principal periodically over the life of the loan, and when the loan matured they were required to make a balloon payment, covering the total amount borrowed. Because it was rare for a farmer to save enough in a short period to fulfill the entire obligation, he or she either asked for an extension from the lender or, more often, negotiated a new loan covering the amount still owed. It was not uncommon for farmers to refinance an original loan three or four times before finally paying it off. Farmers were accustomed to this system, and as long as their property was rising in value they were comfortable with it. It regularly thrust them, however, into a capricious credit market and sometimes placed them in a position where they could not borrow enough to fulfill their mortgage obligation. When that happened, they were in danger of losing their farms.

Determining the maturity of the loan was a matter of negotiation between borrower and lender, in which each was guided by a complex and shifting set of considerations. Longer-term loans provided borrowers with more certainty, relieved them of the necessity of paying commissions and fees frequently, and kept them out of the unpredictable credit market. There were, however, several advantages to shorter-duration loans from the farmer's point of view. The late nineteenth century was a period of deflation, in which prices and wages fluctuated but generally declined. Deflation harms debtors because it compels them to discharge debts with more expensive dollars, in real terms, than the dollars they received when they took out the loan. The shorter the term of the loan, the less harmful the effect of deflation. Deflation also generally reduced interest rates over time. Why saddle yourself for five or six years with an interest rate of 10 percent when you might be able to borrow money at 8 percent two years later? As areas became more densely populated and economically developed, land values rose and interest rates declined. When collateral was more valuable and lending was less risky, lenders would agree to larger loans at lower rates. Why wait for seven years for this to happen when it might happen in two? For their parts, lenders generally preferred not to have their capital tied up for long periods, and mortgage originators profited from the commissions that came with frequent refinancing. On the other hand, if they believed interest rates were likely to fall, lenders tried to lock in rates for longer periods, and they desired to hold on for as long as possible to dependable borrowers with whom they had established a relationship. Whatever loan duration the parties settled on, it was hardly written in stone. Borrowers often asked for loan modifications, and lenders were sometimes willing to renegotiate with valued clients when interest rates fell during the life of the loan.[15]

The amount of the loan was the second arena for contestation and negotiation between lenders and borrowers. Lenders were usually careful to follow three basic rules: make sure that the collateral was worth more than the amount loaned, never lend on property likely to decline in value, and attempt to assure that the borrower had sufficient income to cover the payments. The smaller the amount of the loan, the greater the earning potential of the borrower, and the more valuable the security underlying it, the more likely it would be paid off. Especially when they were arranging loans in recently settled areas and with borrowers with whom they did not have an established relationship, brokers tended to be very prudent, not least because it was difficult to find buyers for questionable loans. The Davenports and J. B. Watkins both limited loans to one-third or less of the value of the property being mortgaged. This was subject to modification, however. Dependable borrowers with a successful track record with a firm might be able to borrow more. And when there were boom times on the plains, as in the early 1880s, lots of mortgage companies with lots of money to lend entered the business. Increased competition among lenders led to more permissive lending standards and lower interest rates, especially when the debenture system facilitated the hiding of low-quality mortgages.[16]

Borrowers frequently wanted or needed more money than a lender was willing to lend. To bridge the gap between what they received and what they desired, they took out additional mortgages, designated "second," "third," "fourth" mortgages, and so on. Because the holder of the first mortgage had priority in a bankruptcy, holders of subsequent mortgages were taking a greater risk. To diminish their risk, lenders virtually always advanced less money to the borrower than the first lender had and, in order to compensate themselves for their risk, they demanded that the borrower pay a higher rate of interest. It was not uncommon for farmers on the plains to have several mortgages, held by several lenders, of varying durations and at different interest rates. Between 1908 and 1914 Thor and Gjertru Birkel mortgaged their homestead at least eight times, in addition to borrowing from Thor's brother. Their story was extraordinary, perhaps, but hardly unique.[17]

The third area of contestation and negotiation between lenders and borrowers was interest rates. Interest rates were determined by global, local, and even personal factors. Overall, the supply of investment capital and the demand for capital were of primary importance. Large and small investors in search of safe and predictable returns provided the capital for mortgage companies operating in the West. As long as the plains economy was growing—as it was in the early 1880s and again after the turn of the century—eastern and European capital flowed in that direction. These capital flows helped hold interest rates in check, but the high demand for money on the part of plains settlers limited the degree to which interest rates might fall. The sharp decline in the western agricultural economy in the late 1880s, and the subsequent national depression beginning in 1893 dramatically diminished eastern and European investment in western mortgages. Western demand diminished as well, but not as rapidly, with the result that rates remained relatively high. Interest rates also varied with the locality. As a general rule, rates were higher in recently settled areas than they were in more established ones. Settlers in newly opened areas needed to borrow, but lenders worried about population turnover, low value of collateral, and a short production record. When residents of such areas could borrow at all, they paid high interest rates. Likewise, because their commercial prospects were less bright, farmers in places not served by railroads paid higher rates than those in areas with railroads. Farmers in established places with longer economic histories and adequate transportation facilities enjoyed the lowest interest rates. In several southeastern Nebraska counties, average mortgage interest rates fell from 11 percent in the 1870s, when the area was newly settled, to 5 percent in 1906. In areas where there was sharp competition for farmers' business, rates tended to fall. And good, dependable farmers with a reputation for prudence and responsibility could borrow more cheaply than could their less respected peers, simply because the risk of lending to them was lower.[18]

However favorable their loans might have been in relative terms, many farmers believed they were badly used by lenders. Interest rates were generally high when considered in the context of the deflationary trend in the overall economy. Ten percent interest on a two-year loan really was closer to 14 percent if prices were declining by 2 percent a year. Lenders pointed out that there was little local capital and that high returns were required to induce outsiders with money to invest in plains mortgages. Farmers, however, tended to believe they were laboring to support parasites who produced nothing tangible but simply lived off the production of others.

Borrowers also complained bitterly about the commissions paid to mortgage companies and to local loan agents for arranging their loans. These fluctuated widely, governed by the supply of investment capital, local demand for loans, competition among lenders and their agents, and guidelines issued by mortgage companies. As a general rule, interest rates and commissions had an inverse relationship; when the former were relatively low, the latter were relatively high, and vice versa. But unscrupulous local agents gouged borrowers. A local agent of eastern lenders charged Old Jules a commission of $220 on a $1,700 loan, nearly 13 percent. In cases such as that, the amount of the commission could easily exceed the cost of interest on the loan.[19]

The borrower was also responsible for paying "an examination fee, clerk's fees for drawing up the mortgage papers, the cost of the abstract, and the fees charged for recording the mortgage in the county records." If the borrower could not cover these fees with cash, they were taken off the top of the loan, along with the hated commissions and usually part of the first year's interest. When all of these deductions had been made, the borrower might see 80 percent or even less of the amount he or she was borrowing. Many borrowers immediately negotiated second mortgages, often with the same lender, in an attempt to bridge the gap.[20]

Lending and Borrowing in a Boom-and-Bust Economy

The lending and borrowing business on the Great Plains was complicated by the reality of the late-nineteenth-century boom-and-bust cycle in the national economy. Dependent on Europeans, especially the British, for investment capital, lacking a central banking system, and without the fiscal and monetary tools twentieth-century Americans counted on to tame the economic cycle, lenders and borrowers stumbled from unpredictable boom to unanticipated bust, depending mainly on prudent behavior and fortuitous timing to see them through.

On the Great Plains the economy was even more volatile than it was in the country as a whole. Plains farmers operated in a market in which prices were set by world supply and demand. They depended on others to store, transport, and market their crops and to provide them with capital and manufactured products. And they were subject to a capricious climate.

Sometimes the factors shaping the economic lives of farmers on the Plains were mostly positive, and a boom resulted. Between the late 1870s and the mid-1880s international commodity prices were relatively good and American exports of grain and meat were generally strong. The weather was uncharacteristically benign, and rainfall in most places was adequate to produce good crop yields. Some boosters argued that a permanent climatic change was taking place due to tree planting or turning of the soil and that the plains would soon be as abundantly watered as the Midwest. Drawn by dreams of riches and the elaborate promotional advertisements of immigration bureaus, railroads, and other boosters, hundreds of thousands of settlers moved onto the plains. Kansas's population increased by 174 percent between 1870 and 1880 and by another 43 percent during the next ten years. In those same periods, Nebraska's population rose by 267 percent and 135 percent. In North and South Dakota, settled later than the states to the south, the eighties witnessed particularly rapid population growth. Between 1880 and 1890, South Dakota's population increased by 256 percent and North Dakota's increased by 416 percent, with most of the growth coming in the first half of the decade. In 1883, at the height of the "Great Dakota Boom," one thousand people per day came to Dakota Territory. Those who depended on the farmers for their livelihoods—the bankers, merchants, lawyers, and railroads—got caught up in the boom psychology and expanded their enterprises in anticipation of a bright future, as did towns and villages that took on municipal debt to create infrastructure suitable for greatness.[21]

The boom meant that there were lots of new settlers with credit needs and lots of older settlers seeking capital to expand or improve their operations. Fortunately for them, at least in the short term, the Great Plains boom caught the imagination of lenders and investors in the East, who were also enjoying a relatively favorable economic environment during the recovery from the Depression of 1873. As early as 1880, J. B. Watkins estimated that forty major lenders were operating in Kansas, putting the money of eastern investors to work. In areas where competition among lenders and their agents was intense, borrowers sometimes benefited. Agents extended lower interest rates and shaved their commissions to attract business, and they sometimes offered more money than they might have in more normal times and offered it to people of questionable creditworthiness. "So keen was the competition between the agents of Eastern loan companies," one observer remembered, that "the mortgager received a sum large enough to make the transaction quite profitable to himself as a speculation." Another noted that mortgage companies "could not . . . loan the money as fast as it came in. The commissions tempted the agents to place high values on the properties." Eager to generate healthy returns on their money and certain that the future of the Great Plains was bright and secure, individual investors took higher risks than they would have taken a few years before and took them for lower rewards. Loans that were too risky for individual investors could be slipped into trust accounts and passed along to debenture holders.[22]

The boom began to flag in 1887, when lower farm prices and a drought raised questions about the inevitability of western greatness. Some settlers who had taken large loans decided "to cash in on their western venture and 'go back to the wife's folks,'" taking the money lent to them and leaving the farm for the loan company. Walking away from the farm and its encumbrances became more common as the boom turned into a bust.[23]

By the fall of 1889 a full-fledged depression gripped the plains. Settlers fled the region, sometimes with their loan money and sometimes without. In northeastern Colorado one homesteader traded his farm for a wagon, and another swapped his quarter section for a steer. Others escaping the region got nothing at all. Western Kansas lost one-fourth of its population between 1888 and 1898. Some counties lost two-thirds of their population. Touring the recently settled Nebraska Sandhills in 1889, Seth Humphrey noted that all but three of the forty-one farms on which his company held mortgages had been abandoned by the mortgagors.[24]

Those who stayed were devastated by the agricultural collapse. "The price of wheat fell one-half," one observer commented, "and those who remained talked hardship and pessimism . . . incessantly." Debtors who remained on the land could not meet their obligations and sought forgiveness, restructuring, or extensions from lenders. Mortgage companies had their own problems. Burdened with nonperforming assets and beset by investors demanding payment, they paid the price for lending practices during the boom that now appeared imprudent. Many went bankrupt, and those that did not survived only by sharply curtailing lending. Lenders foreclosed on delinquent borrowers as a last resort. It is a common misconception, perpetuated at times by American popular culture, that lenders want borrowers to fail because that allows them to acquire the mortgaged property. In fact, foreclosed property represents dead money to a lender. Lenders usually cannot recover their investment and are responsible for taxes and maintenance while attempting to do so. Still, they did not always have an alternative, particularly when they needed liquid capital to meet their own obligations. Within a very short time, credit on the plains was scarce at best and nonexistent at worst. Were it not for the willingness of some lenders to take shares of crops rather than cash in payment of interest, the mortgage market might have ceased to function entirely.[25]

The hard times of the late eighties and early nineties reshaped the political environment of the Great Plains and intensified farmers' anger and resentment towards lenders. Many farmers embraced the program of the Farmers Alliance, which was later transformed into the Populist Party. The Farmers Alliance favored expanding the money supply by monetizing silver. That step would address the problem of deflation, which was rooted in the fact that the supply of gold and gold-backed currency did not keep pace with the level of business activity. Expansion of the money supply would inflate all prices, including farm prices, and make it easier for debtors to meet their obligations. At the state level, debtors expanded on earlier efforts to secure the

Presho, South Dakota, State Bank and Western Investment Company, early twentieth century. South Dakota State Historical Society (SDSHS 00006 02 2). Used by permission.

passage of usury laws, placing a ceiling on interest rates charged by lenders, and of homestead exemptions, which prevented creditors from taking homes in foreclosure proceedings.

These well-intentioned initiatives and laws were only minimally effective and sometimes even counterproductive in a financial system characterized by interregional and international capital flows. Agitation for monetary expansion through coinage of silver disturbed eastern and European investors, who feared being paid in depreciated currency. This fear led some to further limit their western investment, exacerbating the very capital shortage silver coinage was partially designed to reverse. The shortage of capital was also made more acute when some mortgage companies ceased operations in states with onerous usury laws. More commonly, lenders frustrated the purpose of ceilings on interest rates simply by charging higher commissions.[26]

The end of the Great Plains boom and the subsequent economic depression resulted in some changes in the region. The abandonment of many farms and the collapse of real estate values created opportunities for expansion by farmers and ranchers with liquid capital or an ability to borrow, as well as the optimism and courage to take the risk. Most of the mortgage companies either went bankrupt or withdrew from the region, but by the early twentieth century the credit vacuum was being filled by local banks and, especially, by national insurance companies willing to hold farm mortgages in their portfolios. A new boom developed soon after the turn of the century, fueled by a significant improvement in farm prices and adequate moisture over most of the plains. In 1904, Kansas newspaper editor Charles Moreau Harger reported that "the settler who goes into the West to-day with little money can find plenty of assistance in borrowing . . . The farm loan that costs more than six per cent. interest . . . is the exception." This boom lasted until 1920, when another collapse in the agricultural economy brought a new wave of farm bankruptcies and failures of lending institutions. This pattern reappeared most recently in the 1970s and 1980s, when a real estate boom resulting from a rapid rise in grain prices was followed by a collapse in prices and land values and widespread hardship for farmers. Every generation of lenders and borrowers on the plains seems fated to relearn the lesson of the 1880s—that if one hopes to profit from plains real estate one must be nimble or lucky.[27]

4 How They Built Communities

In his 1931 classic, *The Great Plains*, Walter Prescott Webb used a geological metaphor to differentiate institutions on the plains from those farther east. "As one contrasts the civilization of the Great Plains with that of the eastern timberland, one sees what may be called an institutional *fault* . . . At this *fault* the ways of life and living changed. Practically every institution that was carried across it was either broken and remade or else greatly altered."[1]

Webb was an environmental determinist with a penchant for sweeping statements. He believed there were significant differences between life on the plains and life in the places from which most plains settlers came. In terms of their institutions, however, the world they fashioned on the plains was more similar to than different from the worlds they had left. While the settlers "were tantalized by the prospect of building a new world for themselves," in practice they created "a new world that would, in most respects, resemble the world they had left behind."[2]

Settlers carried their traditions, customs, celebrations, and even foodways from the Midwest, Germany, Norway, Russia, and other places to the plains. They recreated neighborhood and kinship networks that closely resembled those in which they had been enmeshed in the places they had left. The institutions they built, such as churches and schools, differed little, if at all, from the religious and educational institutions they had known in their homes.

The economic infrastructure created by people on the Great Plains was also similar to what most had known previously. Plains settlers functioned in a capitalist economic system and strove for commercial success. They cherished private property and accepted a market system that determined the prices and wages they paid and received. They marketed their grain through elevators, patronized lumberyards and implement dealers, and depended on a range of retailers to meet their needs as consumers.

They created local governments based on municipalities, townships, and counties that were identical in form and function to those existing elsewhere. They had the same set of officeholders, elected in the same manner, usually through universal male suffrage—the "institutional fault" of which Webb wrote failing to extend to gender relations. They wrote state constitutions that differed little from those of other states, even going so far as to lift entire passages out of other states' documents and place them in their own. Most plains settlers from Europe were unfamiliar with the details of American governmental institutions and civil subdivisions, but they fit in quickly. Because most were literate and experienced with local self-government, immigrants made the transition to American politics and public life relatively easily.[3]

None of this should come as a great surprise. The people creating a social, economic, and political life on the plains understandably took their models from what they knew. And the very strangeness of the plains put a special emphasis on recreation of the familiar; there was no better way to bring comfort to a strange and frightening environment. But nothing can be recreated exactly. Inevitably there are differences between the new and the old because conditions force modifications or because people disagree or fail to remember all of the details of what was. How the people of the Great Plains constructed and recreated their local communities and economies is the subject of this chapter.

Creating Community Institutions in the Countryside

From the very beginning of settlement, the isolation of people on the Great Plains made the creation of community institutions both difficult and necessary. Visiting North Dakota in 1893, journalist E. V. Smalley wrote that "each family must live mainly to itself . . . shut up in little wooden farmhouses" separated from neighbors "always more than half a mile" away. He found the isolation "bleak and dispiriting," especially in winter. A few years later, Charles Moreau Harger observed in western Kansas that "the nearest neighbor" was "not closer than half a mile, at the best" and that "amusements were few and far between." Other factors complicated the situation. Neighbors sometimes spoke different languages or embraced antagonistic social practices or religious persuasions. Then there was the continual churning of the population, especially in the early years and in homesteading areas, with a quarter or more residents of the locality sometimes turning over from year to year. In challenging economic times, such as the settlement bust of the late eighties and early nineties, many of those who fled were not replaced, leaving a depopulated countryside that intensified the isolation of those who remained. Insofar as community is based in proximity, homogeneity, and stability, constructing institutions should have been a daunting challenge for plains settlers.[4]

But no challenge was sufficient to overcome what Smalley called "the natural gregarious instinct of mankind" on the plains. Before there were tangible social institutions there was sociability, manifested especially in the universal rural practice of visiting. Sometimes visiting simply involved neighbors dropping in or stopping to chat when seeing a farmer working in the field or a wife hanging laundry on the line. But visiting was also a more formal activity, with one family invited to another's house for a hog butchering, supper, or an evening of singing, conversation, and games. Visiting could also involve larger groups that gathered for communal work or for such leisure activities as dances, which were usually attended by entire families. Hattie Lee remembered fondly that "in the old days we went to dances at private homes. The place where we went they would prepare a midnight lunch."[5]

When community institutions and public accommodations were developed, the opportunities for visiting were multiplied. "Inland" stores—so called because they were located in areas not yet served by railroads—quickly followed settlers. They became centers for visiting, in large part because they frequently doubled as post offices. Farm people picking up mail, buying necessary items, or trading eggs, furs, or other products from home would take advantage of a visit to the store to catch up on the local gossip, discuss crops and weather, or debate current events. Other businesses that followed inland stores, such as blacksmith shops, also became venues for sociability, though most that were work-related were patronized exclusively by men.[6]

As neighbors got to know one another, they created more opportunities for sociability, some formal and some informal. Sometimes work was com-

Friends and neighbors gather to raise a barn near Mitchell, South Dakota, 1910. South Dakota State Historical Society (SDSHS 2010 09 17 008). Used by permission.

bined with play, as in threshing, breaking, and husking bees, quilting and feather-striping parties, and house and barn raisings. Events such as quilting parties and hunting contests were gender specific, but many others were communal, including "church suppers, farmers' picnics, . . . strawberry socials, ice cream socials, basket socials . . . roller and ice skating, sleigh rides, toboggan sliding, teas . . . weddings . . . and bob-sled races." It is a wonder that these people, supposedly so lonely and isolated, had time to work.[7]

These activities were manifestations of the underlying kinship and neighborhood networks on which community was founded on the plains and in rural America generally. Local networks were easier to construct when neighbors had kin connections, shared an ethnic or religious identification, or had migrated to the area as part of a group. But newcomers who were not part of a pre-existing network usually became part of one quickly because helpful neighbors were crucial to settlers' survival and success. As Kansas pioneer Roderick Cameron pointed out, "The new settler could be just as helpful to us as we were to him. If we had helped him build his own sod house, as we often did, we could readily get the loan of his plow, or other farm implement, whenever needed."[8]

Neighborhood and kinship networks benefited everyone, but they were created and sustained by women, who played the primary role in producing rural society on the plains. Many activities either revolved around food or included sharing a meal cooked by women. But women were responsible for much more than preparing the food. They planned the events, assigned tasks, and developed and sustained traditions.[9]

The significance of neighborhood networks on the Great Plains exceeded their recreational value. Neighbors such as those in Campbell's Kansas community traded work and shared implements or draft animals. They planted or harvested a neighbor's crop when he or she was ill or injured. They provided help to mothers giving birth and aided in preparing the body for burial when someone in the neighborhood died. They could help tide others over in times of trouble, providing food, clothing, and shelter and even taking in orphaned children. Neighborhood and kinship networks created and maintained a safety net for rural Americans in the nineteenth century, offering many of the benefits we expect the state or social service agencies to provide today. For these reasons, the development of relationships with neighbors was nearly as essential to plains settlers as was breaking the land and constructing a dwelling.

Neighborhood and kinship networks usually preceded the development of more formal social institutions and almost always played a crucial role in their creation and maintenance. Of the social institutions that come to define the rural plains, churches most often came first. Church development was closely tied to ethnicity. Colonies of immigrants undertaking group settlement were often accompanied by ministers. Clergy sometimes recruited and organized the colonists and negotiated with railroads and land companies on their behalf. Such immigrants arrived in their new homes with an essential component of their culture—the church—intact and operational. Moreover, because religion was such a key component of culture, some immigrants became much more devoted to the church in the United States than they had been in their home countries, where church membership was often a matter of birth rather than choice.

Parishioners in front of a Lutheran church near Kearney, Nebraska, ca. 1903. Nebraska State Historical Society (NSHS nbhips 13419). Used by permission.

Immigrants who did not settle in clusters near others from their homelands had a harder time satisfying their religious needs. In some cases they joined an unfamiliar church or conducted services with lay leadership at someone's home. As Jews in an overwhelmingly Christian country, Rachel and Abraham Calof found it especially difficult to observe their faith. When their first son was born, they had to bring a *mohel* in by train to perform the circumcision. The fare and his fee totaled ten dollars, hardly a trivial amount to a family that sometimes had too little money to buy enough coal to last the winter. In later, better days, the Calof home "became the center for all the Jewish holiday celebrations" among the Jews in the neighborhood, but they were never able to build a temple or secure a rabbi to conduct religious services.[10]

American-born settlers on the plains were less likely than immigrants to be accompanied by their ministers. Religion was usually a high priority among the devout, but the resources to build a church and call a minister were not always readily available. The dominant Protestant denominations maintained vigorous home mission programs that helped new communities build churches, and the railroads usually pitched in by providing passes to ministers and by donating land for church buildings, but it was not uncommon for local Presbyterians or Methodists to get by for some years with lay-led "Sunday schools, singing meetings, and other informal . . . activities supplemented by itinerant preachers and camp meetings." These informal services were usually held in settlers' homes or in schoolhouses when they were available. When churches were finally built, the congregations often struggled because their few parishioners were scattered over a large geographical area, a problem that lasted far beyond the early settlement era. Only rarely could a congregation on the plains support its own minister. More commonly, two or more churches shared a minister, who divided his time between them.[11]

Even struggling churches were highly valued by their patrons. They provided the essential services of explanation, celebration, and consolation. At their best, they embodied caring and nurturing communities on which parishioners could depend in times of trouble. They were also recreational centers, hosting Sunday schools, special services, dinners, and socials and sponsoring women's circles and youth groups. At their most basic level, they offered something to do, which explains why members got together on Sundays for hymn singing and devotions even when the minister was not present. As Faye Lewis remembered of her settler family in southern South Dakota, the church "service was the one break in a rather lonely week. It was going someplace after a week of staying at home, a time of social contacts after a week barren of them, talking to people outside one's family, visiting with friends." Churches were religious institutions, but to some the religious component was secondary.[12]

Men usually governed churches, and female preachers were unheard of in most denominations. But the women in the congregations sustained most of the churches—planning activities, preparing food, and raising money for special purposes. In this, as in other areas of plains life, women took the responsibility for maintaining community institutions.

Churches were essential to the *communities* they embodied but were not invariably positive to the larger *community* of which they were a part. Although congenial denominations such as the Congregationalists and the Presbyterians frequently cooperated, doctrinal differences and disputes sometimes divided neighbor from neighbor. When churches attempted to impose moral standards on others, especially when they battled alcohol, they created discord in rural neighborhoods. Ethnic churches were foundations of strong ethnic identities and were to some degree dependent on the survival of those identities for their institutional health. The reinforcement of ethnic identities unquestionably enhanced the comfort of people in a strange land, but it also helped impede the assimilation of immigrants to American culture and separated them from others. Schools, on the other hand, were community-wide institutions that could bridge religious and ethnic divides. Creating them, though, was sometimes more difficult than creating churches.

Government policies provided support for public education on the Great Plains. When federal lands were surveyed, sections 16 and 36 of each township were set aside for support of public schools. When territories were transformed into states, their constitutions provided for free public education, and state laws required schools when the population of children in a particular area reached a designated threshold. Statehood also brought a grant of federal land to each state, some of which was designated for the support of public education. The problem was that financial inducements from the federal government and the states, well-meaning though they were, covered only about 20 percent of the cost of providing public education. Schools on the plains had to be supported mainly by local property taxes, as they were in the rest of the country.[13]

Most plains settlers were literate in some language. They had enjoyed the benefits of education when they were young, and they wanted the same for their children. But too often they lacked the resources to act on those generous and positive impulses. Settlers usually had little extra cash, and they poured what they had into their farms. Those farms demanded most of their attention and the labor of their children, and it was sometimes years before they turned a profit. In areas where homesteading predominated, there was little property tax revenue to be had because settlers paid none until they proved up. Everywhere farmers worried that the value of their land would decline and potential purchasers would be dissuaded from buying if they encumbered the land with heavy local taxes. Adding to this was the high level of transience in many areas, especially in the early years, and the substantial number of single settlers, usually less committed to education than were their neighbors with children.[14]

Parents determined to provide education for their children sometimes started "subscription schools," patrons of which paid tuition for their children or conducted "various benefits and socials" to raise money. Alternatively, an enterprising teacher might open a school and publish a fee schedule. Subscription schools were often conducted in private homes. Nina Farley Wishek taught school in a one-room house she shared with the resident family and sometimes their farm animals. Her seven or eight students spoke no English, and she did not speak their native German.[15]

When the financial position of settlers improved, most communities began taxing themselves to build public schools. That process often led to conflicts. Neighbors struggled over where the school should be located. Location became a major issue because settlers perceived that a nearby school would enhance their property values and be convenient for their children. Settlers also argued over the length of the term. Farmers wary of the cost of education and those who depended heavily on the labor of their children favored as short a term as the law allowed. Differences also arose over whether parents or the school district would be responsible for buying books and whether standard books should be required. Even when the community committed itself to creating a school, the process was neither smooth nor painless. In Roderick Cameron's neighborhood, the creation of a school was delayed for a year by a dispute over location and the lack of money to hire a teacher. The following year the school closed after three months when families could not pay a one-dollar monthly fee.[16]

The schools that were eventually created reflected the marginal economic situation of most settler communities. The typical school was a one-room, wooden box, poorly insulated, and heated by a coal or wood stove in the center. Children near the stove roasted, while those on the periphery froze. Some schools lacked privies; most had one shared by boys and girls. Children drank out of a water bucket using a common dipper. Grades 1 through 8 were taught together through the recitation method, which stressed rote memorization. The curriculum emphasized the three Rs and was heavy on patriotic history and civics. Attendance was low when weather was threatening or when farm work kept the bigger boys at home. Local control was the rule in one-room school districts. There might be requirements at the state level, supposedly enforced by county superintendents, but as a practical matter the local board hired and fired teachers, determined the language of instruction, expended funds on the physical plant, and sometimes even directed what would be taught and how.[17]

By the time white settlement on the plains began in earnest, the teaching force was predominantly female, and it became more so over time. The preference of local boards for female teachers did not indicate a commitment to female equality. It was simply cheaper to hire women than men, though boards sometimes argued that younger children learned better under women, with their supposed natural nurturing skills. The typical one-room schoolteacher was in her late teens or early twenties—sometimes younger than her older students. She had a one-room school education and little more. Those who stayed in the profession longer than a year usually upgraded their skills by attending teachers' institutes or taking summer classes in normal schools, two-year institutions that specialized in training teachers. They were responsible for teaching as many as forty students, ranging in age from five to twenty or older and varying widely in ability. They were also expected to organize spelling matches and declamations to demonstrate their abilities to the community and to plan and execute the annual Christmas program and the year-end picnic and program. Teachers doubled as janitors, cleaning the school every day, drawing the water, building the fire, and carrying out the ashes.

One of the major challenges confronting teachers was maintaining discipline, especially among adolescent boys. In Roderick Cameron's district, some of the older boys tied the teacher to a fencepost and left her there all day. He noted laconically that "treatment like that compelled her resignation." Teachers were required to set an example for their students and the community. This meant "no drinking, no dancing or card-playing where the community attitude was against it, no gallivantin' around, no slang . . . we wore ruffled thing-a-mabobs to conceal our maidenly forms. Neither did we show our legs." Sometimes prohibited from dating, teachers were dismissed if they married. For all of this they made about two hundred dollars for about twenty-five weeks' work, plus room and board, which they took at patrons' homes. It was a hard job for meager pay but preferable to work as housemaids

Teachers and students at a one-room school, West Union, Custer County, Nebraska, ca. 1887. Note the range in ages of the students. Nebraska State Historical Society (NSHS nbhips 13610). Used by permission.

or farm laborers, the only other positions available for single women in many rural communities on the plains.[18]

Whatever weaknesses or strengths country schools had as educational institutions, their significance as social and recreational venues was unmatched. In contrast to churches, which sometimes divided rural people, schools brought them together. The school belonged to the entire community and provided a physical space that all could use and enjoy. Some of the social and recreational opportunities offered by schools were directly related to their educational enterprise. Parents and community residents generally trooped in for spelling matches, declamations, and holiday programs. In the winter, schoolhouses often hosted "literaries," community variety programs featuring readings, recitations, music, debates, and short plays. Traveling politicians, orators, and performers used the schoolhouse, which was usually the only local structure that could accommodate any kind of crowd. As a sizable public space, the schoolhouse also provided a meeting place for farmers' and homemakers' clubs, local political organizations, and other groups. Those driving along back roads in the Great Plains today observe many one-room schools that have been converted into township halls that continue to host public meetings and events.[19]

Churches and schools were the most visible representations of the rural community on the plains, but they were not the only ones. In communities with energetic, progressive leadership, a visible institutional infrastructure matched the vital neighborhood networks that underpinned it. In 1882, Henry Martin of Bradley, Kansas, bragged of his "flourishing community, in which one church building had been completed, our public roads laid and maintained by our own labors, and a school bond passed." Eight years later he and "a group of our most wide awake and enterprising farmers" had founded a "Farmers' Institute . . . for the purpose of increasing our knowledge of modern farming methods." Elsewhere farmers formed chapters of the Farmers Alliance and later the Populist Party to protest their economic disadvantages. Others created cooperatives to handle their grain, market their cattle, or turn their milk into cheese. While these were primarily economic and political institutions, they all included a significant social and recreational component, and all were dependent on the neighborhood networks, created and sustained by women, that ultimately underlay virtually all group activities on the rural plains.[20]

Town Building

Businesses serving settlers' needs developed along with settlement. Soon after the first few farmers appeared in an area, a storekeeper would usually set up shop, and he would likely get one of the post offices that were liberally distributed. As the population increased an inland town would often grow up around the general store and post office, perhaps including a blacksmith shop, a livery stable, a mill, and other businesses, along with dwellings for the proprietors and their families. But development of a fully integrated and elaborated economic infrastructure did not usually come until the railroad arrived.[21]

In the East and the older Middle West, railroads generally conformed to the urban patterns that already existed. Cities such as Cincinnati or Saint Louis, on the Ohio-Mississippi River system, or Cleveland and Buffalo, on the Great Lakes–Erie Canal system, were already established agricultural marketing and processing centers when the railroads appeared. The railroads could profit from transporting agricultural products and perhaps influence urban growth, but they could not reshape the established pattern.

On the plains, where virtually no urban development had occurred before the railroads entered the picture, the situation was vastly different. Through their decisions, the railroads helped determine which places would become major centers of agricultural marketing and processing, and such cities as Omaha, Minneapolis, and Kansas City were the beneficiaries of those decisions. On the local level, the railroads determined which places would prosper, which would languish, and which would vanish.

The power of the railroads over economic life and death on the plains was due to the fact that there was no other viable transportation alternative. With the exception of a very few areas, water transportation was not feasible on the plains, and the movement of commodities by wagon was prohibitively expensive. Most farmers without access to rail transportation either produced livestock, which could walk itself to market, or spent inordinate amounts of time marketing crops at a distance. Before the railroad came, farmers in Wayne Township, Kansas, grew corn, which they fed to livestock. They either drove their livestock to market themselves or sold it to independent drovers. When rail transportation became available, they switched to wheat, a less labor-intensive crop that could now be moved to market easily. Before the Great Northern ran a branch line through three nearby settlements in 1911, North Dakota farmer Anders Svendsbye hauled his grain by wagon to elevators twenty miles distant. His wagon box held fifty to sixty bushels and the round trip took two days. Marketing the produce of just forty acres could take him a month, assuming that the roads were passable. The railroad's entrance into a farming region thus reshaped production, facilitated marketing, and raised property values. It brought other benefits as well. The telegraph, an essential tool for railroad operation, improved communications for all residents.[22]

The significant transportation and communication benefits brought by the railroads meant that farmers were "usually overjoyed when the first tracks reached their hitherto isolated tracts of land." But they soon realized that the railroad was not an unmixed blessing. They found themselves dependent on companies that could, and often did, charge whatever the traffic would bear for their services. Other irritants included the deaths of livestock on the tracks, the prairie fires caused by locomotive sparks, and increasing numbers of transients. Nor did the railroads necessarily banish isolation. In the winter of 1879–80, Mitchell, South Dakota, was cut off from the outside world for sixteen weeks when heavy snows shut down the line. Three years later snow closed the Northern Pacific line from Fargo to Lisbon, in northern Dakota Territory, for three months. With relatively little grain to move at that time of the year, the railroads had no incentive to clear the tracks. When the railroad stopped operations, those who had come to depend on it suffered, as did the Ingalls family, which ran out of kerosene, coal, meat, and flour during the "long winter."[23]

The railroads' basic purpose was not to end farmers' isolation or enrich their lives. When railroads pushed into new territories, their goal was to effectively and efficiently exploit the available traffic and to encourage more. This involved developing towns and placing them close enough together—usually eight to ten miles apart—to fully cover the agricultural service area but far enough apart that they did not cannibalize one another's business.

In choosing their routes, railroads seldom responded to independent townsite promoters or to existing inland towns desiring rail service. Instead, they created their own towns, either through development companies owned by or subsidiary to them or through townsite promoters with whom they partnered. Railroads could easily create their own towns because their executives knew where the railroad was going and independent speculators did not. Once the decision determining the route and the location of townsites was made, the railroad's town development agents or partners would quietly go to the area and purchase the townsite, if it was not already part of the railroad's land grant. Then they would prepare a plat of the town (a sort of blueprint complete with streets and lots), name it, and file the plat with the county. Inland towns that were passed by were doomed because they not only lacked a railroad but would be competing with a new town that had one. Merchants in inland towns frequently relocated to the new townsite, lock, stock, and barrel, sometimes moving their buildings as well. That was what Roderick Cameron's father did: "The railroad having refused to come to *us*, our only response was to move to *it*."[24]

Railroad plats followed several standard forms, but the most popular style on the plains was the "T-town." In a T-town, the railroad tracks formed the horizontal bar of the T. On one side, along with sidings to facilitate loading and unloading, were grain elevators, lumberyards, and implement dealers. On the other side, along what was usually called "Railroad" or "Front" street, were the depot and the freight house. The vertical bar of the T was "Main Street," the retail and financial center of the town. Perpendicular and parallel to Main, in a rigid grid pattern, were residential streets.[25]

The main purpose of these towns was to collect agricultural produce in an effective manner, but railroads also profited from the sale of lots. The pattern and practices of townsite development had been set shortly after the Civil War by Grenville Dodge, head of the Town Lot Department of the Union Pacific, and were generally followed by other railroads on the plains. Once a town was platted, townsite agents, usually working on commission, moved aggressively to attract businesses to occupy Main Street lots. As was the case with railroad advertising and sales programs aimed at farmers, townsite promotion served to familiarize potential townspeople with the region being promoted while emphasizing its promise, often in extravagant terms. Agents also offered credit. A down payment of one-third was usually required, with the balance to be paid over two years. Early comers often received a discount on town lots, in order to get the town up and running quickly. Developers usually withheld at least half of the lots from sale, hoping that rising prospects for the town would inflate their value, and they donated lots for churches, schools, public buildings, and other amenities that would enhance property values.[26]

Townsite developers frequently sold lots sight unseen to distant buyers, but many purchasers were merchants from nearby towns looking to expand or to relocate in a more promising venue. When the Milwaukee railroad opened the town of Presho, South Dakota, in 1905, it brought potential buyers in on a special train and began auctioning lots. By the end of the day, sixty lots had been sold, forty along the "Main Street" that had been marked out on the prairie. New towns were full of promise, and "the positive mood was contagious." Visiting Gypsum City, Kansas, in 1887, government pension examiner Leslie Snow noted that "this city (?) . . . Has a bank, hotel, and brass-band . . . Pigs in the street. Citizens say it is to be the metropolis of the world." New themselves and often surrounded by unsettled land, infant towns tended to attract the sorts of entrepreneurs who lived on real estate booms. Roderick Cameron wrote of Kirwan, Kansas, in 1878 that it had "about four hundred people. Nearly every other person seemed to be in the real estate business. Estate agents and law offices seemed to be located on practically every vacant lot." Visiting the new town of Aberdeen, South Dakota, a few years later, Seth Humphrey complained of "too many cheap lawyers and swarms of land men . . . Loan sharks galore . . . ready to finance the settlers."[27]

Townsite promoters welcomed all sorts of businesses to new communities, but they considered some to be necessary. Grenville Dodge recognized that without lumber dealers "it was difficult to sell any lots on a townsite." David Tallman, who handled town development for the Great Northern early in the twentieth century, believed that a successful town needed "three to five lumberyards, one to two banks, two to three general stores, one or two hardware stores and farm machinery dealers, plus . . . a single drugstore, hotel, newspaper, butcher, restaurant, and livery stable." If an essential component was missing or underrepresented, townsite developers would heavily discount lots as an inducement or even give them away to incoming businesspeople.[28]

Everybody involved anticipated making money in the new towns. Grain elevators and lumberyards were usually parts of chains that already had a mutually profitable relationship with the railroads. Smaller merchants hoped to prosper as well, whether from their own enterprises, from real estate, or from farming, which many pursued as a sideline. Small-scale speculators were sometimes disappointed, but townsite developers virtually always profited. In towns that exceeded their expectations, they benefited mightily from real estate sales. In towns that did not match expectations or in which many buyers did not keep up their payments, the promoters dumped their unsold lots quickly and moved on. Even in those places, though, they made money, because their costs had been trivial relative to their profits.[29]

Townsite development was just one component of the economic significance of the railroads to communities on the plains. Railroad decisions to run the line here or there, to make this or that town a division point, or to locate shops in one place or another meant bright futures for some and limited prospects for others. Towns without a railroad labored to attract one, and towns with one worked to attract another in the hope that the competition would lower freight rates. In 1886 the towns of Morris County, Kansas, paid the Rock Island $73,000 to locate a line there that would compete with the Missouri Pacific. Later, Council Grove offered the Santa Fe $70,000 to extend a line there. Overall, in 1887–88 municipal governments in Kansas offered railroads subsidies totaling $8.5 million. Railroads were hardly the only contributors to economic development that towns on the plains tried to acquire. They struggled mightily for public institutions such as prisons and insane asylums, competed vigorously—and sometimes violently—for the designation as county seat, and sought to attract colleges and other facilities that might boost local employment, business, and reputation. But nothing was more beneficial to a community's prospects than a railroad line or a second or third one, which explains the extravagant expenditures towns were willing to make.[30]

People on the Great Plains regarded the railroads rather as present-day Americans regard the airlines: they were absolutely necessary, but their practices were detestable and their service was poor. Farmers who had been excited to see the railroads come soon became disenchanted with them. Luna Kellie argued bitterly that "the best . . . years of our lives were given to enrich the [railroad] . . . They . . . cleared annually more from our toil than had been wrung in the old times from the colored slaves." Enraged by high rates, indifferent service, the corruption of state and local governments, and the railroads' generally imperious attitude, late-nineteenth-century farmers joined the Farmers Alliance and later the Populist Party. But farmers were hardly the only people who had grievances against the railroads. Shippers in towns complained of high rates and favoritism to others, and elevators grumbled that sidings were poorly maintained and insufficient cars were provided to move the harvest. In self-defense, the railroads pointed to their heavy debt loads and high fixed costs—factors that frequently drove them into receivership—and to the expense of operating on the plains, where too often they were compelled to move empty cars to the West in order to move produce to the East. There was truth in these arguments, but they were hardly convincing to those who were victims of the railroads' "governing principle . . . to 'charge what the traffic will bear.' "[31]

Their indispensability to the grain trade meant that line elevators were the most significant businesses in country towns, and most towns had two or more. There were a few independent or farmer-owned grain elevators, but most were owned by large firms that controlled forty or more elevators. These firms worked closely with the railroads as a matter of necessity. They needed to be close to the tracks, so they sought long-term leases on the railroad right-of-way and depended on the railroad laying side tracks and building platforms to facilitate their loading of cars. Frictions arose between the railroads and the elevator companies from time to time, especially over car availability, but their interdependence required that they generally maintain a close working relationship. Further underpinning that relationship was the

A farmer delivers a load of grain to the Eclipse Elevator, South Haven, Kansas, about 1890. Kansas Historical Society (KHS 226848). Used by permission.

fact that elevator companies commonly confined their operations to a single railroad.[32]

Elevators sometimes diversified their businesses by selling seed and fertilizer, but their main function was the handling of wheat and other grains. They were the first step in the process of turning grain into an international commodity. At harvest, farmers delivered their grain to one of the local elevators. There it would be weighed and graded, and the farmer would be paid, based on a schedule provided by one of the grain exchanges located in the major milling and exporting cities. The elevator then marketed through commission merchants, who arranged for the sale and delivery of the grain in Kansas City, Chicago, Minneapolis, or some other grain terminal. There it was either ground into flour—as it was in Minneapolis, the nation's leading milling center—or exported, primarily to Europe.

The elevators' profits were largely determined by the volume of grain they handled. Regardless of the price of wheat in any given harvest season, they charged a flat fee on each bushel they received. Farmers complained, with some justification, that the line elevators fattened their wallets at the farmers' expense in several ways. Farmers believed that elevators short-weighed grain or downgraded it, declaring number 1 wheat to be less-valuable number 2, for example. If the elevator then sold it as number 1, it claimed a profit that rightfully belonged to the farmer. Farmers also accused the elevators of excessive dockage—the levying of a price penalty based on the amount of dirt, weed seed, and broken kernels in the grain. These practices might shift only a few cents a bushel from the farmer to the elevator, but when wheat was selling for forty or fifty cents a bushel, as it sometimes was, a few cents could be the difference between making the mortgage payment and not. For the elevator, which handled tens of thousands of bushels per year, a few cents a unit could make the difference between a good and a very good year.

Lumberyards, like grain elevators, were usually parts of chains, some owned by wholesalers and some directly owned by Great Lakes lumbermen. They, too, needed a lease on the railroad right-of-way and siding to allow the convenient unloading of lumber. As was the case with the elevators, they maintained a cordial relationship with the railroad. But the lumber business, unlike the grain trade, was a boom-and-bust enterprise. When a new town was getting started, the demand for lumber seemed insatiable. New residents were building homes and businesses; new settlers, brought by the railroad, were throwing shacks up; and more established settlers seized the opportunity to rebuild with milled lumber, which had usually been too far distant and too expensive before the appearance of the railroad. But after the building boom was over and everyone who wanted or could afford one had a new house or store or outbuilding, the demand for lumber collapsed and usually remained anemic for several years. The result was commonly a shake out, with some yards moving and others going broke. Those that survived depended on shrewd management and frequently on diversification, especially by selling coal and sometimes hardware, ice, and other items.[33]

Businesses serving producers were essential components of the towns the railroads brought to life. Virtually every town had at least one implement dealer. Farmers on the plains coveted labor-saving machinery such as riding plows, seed drills, and reaper-binders, particularly if they were growing wheat, which was especially amenable to mechanization. They complained of the prices—a riding plow cost $60 and a binder cost $250 in 1878—but the implement business was not a monopoly in the sense the railroad often was. Many towns had two or more dealers, and most dealers carried more than one line. The cost of transporting implements to the plains was high, but the basic problem was that in the early years of settlement farmers had difficulty generating enough income to meet their need for tools that would allow them to generate more income. Blacksmiths were also essential in new communities. Not only did they shoe horses, they also repaired tools and fashioned nails, plowshares, and dozens of other necessary items.[34]

The most common businesses were general stores. These establishments carried a wide variety of goods, including groceries, clothing, fabric, notions, boots and shoes, hardware, patent medicines, sporting goods, and books. The mercantile business was highly competitive, and it became more so over time. In addition to competing with the other stores in the town or the locality, merchants faced competition from peddlers, who carried their goods directly to the farmer's door, sometimes on their own backs. Free of the storeowners' overhead, peddlers could frequently undersell them in the limited range of goods peddlers offered. Another challenge came when farmers bypassed local merchants to shop in larger towns with greater variety and better prices. This problem continues to plague small-town merchants struggling to survive. After the turn of the century, local merchants faced increasing competition from such mail-order houses as Montgomery Ward and Sears, Roebuck and from J C Penney and other chains that located in rural trade centers. Farmers ordered goods advertised in mail-order catalogs and received them through railway express and parcel post, which was instituted in 1913. Their volume business allowed the mail-order firms to undersell town merchants, even when transportation costs were factored in. J C Penney and other chains offered greater variety at better prices because they could buy in bulk and refused to extend credit to customers. The cost of doing business for local merchants was high. They had to maintain a physical store, and they could not afford the volume purchases and carload shipping that would have kept their costs down.[35]

Small-town merchants coped with their competitive challenges in a variety of ways. Those few with extra capital opened branches in nearby towns. With more outlets they could buy in bulk, lowering their unit costs. Another strategy involved diversifying by adding new enterprises to their core businesses. It was common to see general stores that were also butcher shops, undertakers, implement dealers, or something else. General stores also attracted business by offering favorable terms of trade. They took such items as animal pelts and butter and eggs in exchange for store goods. Sometimes, to hold a customer's business, the storekeeper offered him or her more than the bartered items were worth. Virtually all merchants offered credit, which was their main advantage over peddlers, mail-order houses, and chains. The problem was that they were unlikely to be paid until harvest, and sometimes not even then. In the meantime, they depended on credit from wholesalers or local bankers, who dunned them for money and threatened to withhold further goods from them. The basic problem confronting small-town merchants on the plains was that there were too many of them for all—or even most—to prosper, especially in areas with a sparse settler population with very little discretionary income.[36]

Larger and more promising towns attracted a wider range of businesses than the minimum necessary for a marketplace. Bakeries, confectionaries, jewelry stores, theaters, laundries, dressmakers, and millenaries were common in towns of five hundred or more. Dressmakers and millenaries were especially important because there were female-owned and -operated enterprises at a time when there were relatively few acceptable business opportunities for women. These shops also provided places where town and farm women could meet and socialize. They were a counterweight to the saloons, livery stables, pool halls, and blacksmith shops that were exclusively male domains.[37]

Economic necessity was the basic reason for the existence of towns on the

Interior of the well-stocked Harthorn General Store, De Smet, South Dakota. South Dakota State Historical Society (SDSHS 2009 07 15 004). Used by permission.

plains; central places were required to gather the produce of the country efficiently and to provide services to farm families as producers and consumers. But towns developed their own vibrant community institutions. A perusal of annual events in almost any plains town at the turn of the century reveals a striking number and variety of social and recreational institutions and activities. In the towns as in the countryside, the churches were centers of social organization and recreation, providing networks of auxiliaries and circles, youth groups, Sunday schools, services at least twice a week, and outlets for philanthropic and charitable impulses. Schools were also significant social institutions that hosted a wide range of recreational activities beyond their educational mission. Because of the greater concentration of population in towns, schools were usually graded and employed more experienced and professional teachers than did those in the countryside. When public high schools began to appear on the plains in the late nineteenth century, they were located in the larger towns; rural parents who wished their children to attend had to board them in town and often pay tuition.

These formal institutions were supported and sustained mainly by middle-class women, the wives of the merchants, bankers, doctors, lawyers, and elevator managers who composed the town's elite. Elite women were less likely to have the heavy labor demands burdening farm women. Moreover, they frequently had live-in maids, drawn from the nearby rural area. Their relative freedom from onerous labor not only provided them with the leisure to support civilizing institutions but also allowed them to create a dense network of clubs and organizations. Many clubs were local, but virtually every town of at least five hundred had a chapter of the Women's Christian Temperance Union, the most important moral reform organization on the plains. The W.C.T.U. championed prohibition but also battled against cigarettes, pool halls, immoral literature, prostitution, sexual abuse of children, wife-beating, and other practices and behaviors it considered degrading to the moral tone of the community.[38]

Men as well as women were involved in nurturing community in towns on the Great Plains. Men's activities frequently had a commercial purpose. They organized celebrations on patriotic holidays or at harvest at least in part to bring rural consumers in. Of these celebrations the most important, by far, was the Fourth of July, which commonly featured oratory, band concerts, food, fireworks, and sometimes a fair or a circus. For Kansas farm girl Luna Warner, the Independence Day celebration in nearby Cawker City was a day-long event involving "singing, speaking, and reading," baseball, sack races, dinner, dancing "all the afternoon," and "after the lanterns were lighted a minstrel performance."[39]

In addition to organizing community celebrations, movers and shakers formed commercial clubs to boost the town and to unify the business community. They also created a variety of other organizations that combined social, recreational, philanthropic, and sometimes economic and political purposes. Towns with more than a few hundred people usually had at least one fraternal organization—most often a Masonic lodge—and a chapter of the Grand Army of the Republic, the Union veterans' organization. Because lodges usually had female counterparts or auxiliaries, they provided recreation for women as well as men. Even very small towns usually had a band that practiced, paraded, and gave free concerts and a baseball team that

Towns were important centers for recreation on the Great Plains, as this community celebration in Redfield, South Dakota, demonstrates. South Dakota State Historical Society (SDSHS 2011 07 11 313). Used by permission.

challenged nines from neighboring communities. In addition, commercial recreation was often available, provided by visiting lecturers, Chautauqua performers, theatrical companies, and circus troupes. Plains towns did not offer the excitement of city life, but anyone who couldn't find anything to do wasn't looking very hard.[40]

The relationship between town and country on the plains was sometimes "cordial," but farm families and town families moved in separate, albeit partially overlapping, circles. They were mutually dependent, but they were separated by "distance, ethnic differences, and cultural background." As a practical matter, distance made it difficult for farm people to participate regularly in the life of the town, even when they had the time and money to do so. People in the towns were more likely than country people to be older-stock Americans, and farm people were more likely to be immigrants. This problem was not insurmountable, and farmers and merchants learned enough of each other's language to maintain commercial relationships, but most people on the plains, like people everywhere, tended to be more comfortable with people like themselves. Farmers also sometimes distrusted townspeople, especially those in larger and more sophisticated places who tended to put on cosmopolitan airs. They sensed that people in town viewed them in a condescending way as rubes or hicks. By contrast, they tended to see themselves as producers of the basic commodities that sustained human life and to see the merchants, bankers, and elevator men in town as parasites who lived by manipulating farmers. These divisions were usually submerged, but they came to the surface during the Populist movement, when townspeople assumed a much more conservative political position than that of farmers, and later during World War I, when German and Scandinavian farmers were more likely than villagers to oppose the war. Despite these divisions, however, people in town and country were united in a common quest—to wring a decent living out of a difficult environment and to create a satisfying community.[41]

5 How the Plains Matured

WRITING IN *Collier's* magazine in 1909, Sarah Comstock dramatized the metamorphosis the Kansas farmer's wife had undergone over the previous quarter century. In place of "the shake cabin or the soddy," she lived in "a twelve-rooms-and-bath" house. She had "chopped up grocery box furniture for kindling and bought cut-glass cabinets and antique mahogany highboys from Grand Rapids." And she had "relegated the . . . buggy to the barn and substituted an automobile." The Kansas farm wife was no longer oppressed by hardship. To the contrary, she was now beset by excessive affluence, surrounded by too many things and burdened with too many social activities and obligations. She had overcome the challenges of the settlement period and now enjoyed a life of abundance in a mature, stable community.[1]

In Meade County, South Dakota, the year after Comstock's article appeared, William Roth was embarking on the journey the Kansas farm wife and her family had apparently completed. He purchased a relinquishment for $420 and began the process of making a farm, breaking land, securing wood and water, and doing all of the other things the Kansans had presumably done twenty-five years before.[2]

These experiences illustrate the reality that settlement of the Great Plains was not an *event* but a *process*. Plains settlement began in earnest after the Civil War and continued until the United States entered World War I and even beyond. The settlement process was not smooth. It was characterized by booms and busts, surges and recessions. The late 1870s and early 1880s witnessed a remarkable boom in the eastern and central plains that ended abruptly with plummeting yields and prices beginning in the mid-eighties. After 1900 a new boom developed on the central plains and spread westward. If entries under the Homestead Act are any indication, the second boom was more impressive than the first. Between 1901 and 1913 twice as many homestead claims were filed per year than in the thirty-eight previous years, peaking in 1913.[3]

To suggest that settlement was a process might imply that it culminated in something, that the region achieved some sort of equilibrium, that all of the sod houses and buggies gave way to elegant frame structures and automobiles and that all of the settlers enjoyed success. This was not the case because real human societies never achieve true equilibrium for very long. They might attain a high level of economic and social stability of a sort for a time, but there are always forces beneath the surface or built into the system itself that put it in peril. Climate, the structure of the economy, and human aspiration conspired to create instability on the Great Plains.

The Golden Age of Agriculture and the New Plains Boom

The surge of settlement activity on the central and western plains after 1900 was fueled by several factors. Agricultural commodity prices worldwide rose faster than the general price level between 1900 and 1920, making farmers relatively prosperous and bringing about the "Golden Age of Agriculture." Land prices rose along with commodity prices, promising farmers significant capital gains on their initial investments. The railroads stimulated the boom by resuming the construction of branch lines throughout the plains after the depression of the 1890s eased. Lenders, including new ones such as deep-pocketed insurance companies, provided ample credit. And the early twentieth century was characterized by benign weather and ample moisture on the western plains, increasing yields and generating optimism among farmers.[4]

Contemporaries frequently contrasted post-1900 pioneers, who mostly settled on the western plains, with their predecessors who had settled in the eastern part of the region a quarter century earlier. Those who came later seemed to have it easier. As Kansas newspaper editor Charles Moreau Harger put it in 1904, "The settler of today . . . rides in a comfortable railway coach to the very locality where he is to reside. He drives out to his new home in a buggy, instead of toiling over the way in a 'prairie schooner.' He is close to market and to well-appointed stores. Perhaps he has free rural delivery of mail." Farm work was now less taxing. Massive steam-powered tractors broke sod easier and quicker than could plowmen with oxen. Small, portable gasoline engines lightened many tasks on the farm and in the home. And steam drilling made well digging a less onerous task than it had been for earlier settlers.[5]

It was not just the settlement experience that was different. The new settlers seemed different from their predecessors as well. More of them were single, and many more of them were women. In northeastern Colorado 42 percent of homestead filers after 1900 were single, as opposed to 21 percent before, and 18 percent were women. They had claims in the countryside, but many of them worked in town, living on their claims the bare minimum stipulated by land legislation, if that. As clerks, teachers, and milliners, the new pioneers appeared frivolous to more experienced farmers. That impression was supported by the fact that many seemed to know little about agriculture. Regarding the western Dakotas and Montana, the consensus estimate was that "half or more of the first-time homesteaders . . . had no relevant farming background."[6]

A lack of farming experience was a serious impediment if settlers wanted to farm, but many on the western plains were considered speculators who "counted on the efforts of their neighbors, rather than upon the value of any improvements they might make, to drive up the value of their claims." In those cases, a rising tide lifted all boats, even those piloted by amateurs. Single female homesteaders' motives were especially subject to suspicion, particularly on the part of those who could not believe that women might actually be serious farmers. "First and foremost these women expected a financial return on their investments . . . For . . . the majority, the land represented a speculative venture." When as far as 200 miles west of the Missouri River a 160-acre farm could triple or quadruple in value between 1900 and 1910, a claim the holder intended to relinquish or commute held unquestionable speculative appeal.[7]

There were clear contrasts between those who settled on the western plains after 1900 and their predecessors. As Charles Moreau Harger observed, technological advances eased the tasks of homemaking and farm making. And many women had higher expectations and greater self-confidence than their mothers had, a reality reflected in their willingness to file on homesteads. But other differences emphasized by contemporaries were more apparent than real and perhaps demonstrated nostalgia and the very human assumption that the present is always inferior to the past. In northeastern Colorado, for example, 80 percent of homestead claimants after 1900 had farming experience, mostly on the Great Plains. Rather than being feckless and frivolous, these folks seemed relatively serious. Over half of them—including over half of female claimants—proved up, a proportion in line with or even slightly higher than that in most places on the central and eastern plains.[8]

There was certainly substantial population turnover on the western plains, but that was a reality throughout the region over the entire settlement period. Living in eastern North Dakota in the 1870s, Mary Dodge Woodward noted that "nobody keeps track of his neighbors here. People come and go; families move in and out, and nobody asks whence they came or whither they go." In Kansas, Nebraska, and Dakota Territory, "only 39 percent of farmers . . . could be found in the same township in 1870 and 1880." Striking as that percentage is, it actually *understates* population churning by failing to count settlers who came after the 1870 census and left before the 1880 count.[9]

High population turnover on the Great Plains was due to several factors. Some people came with the intention of moving on. Settlers looking to make a quick profit and push farther west had been a prominent feature of frontier life ever since independence. The Homestead Act allowed people to acquire a potentially valuable resource for virtually nothing, adding to the numbers of those looking to flip their properties. Others simply found plains farming too onerous or plains living too harsh. Farming on the plains was different from farming farther east, and people could become discouraged or go broke before learning how to do it. But failure did not always result from individual deficiencies. Agricultural success meant commercial success, and "the margin between success and failure was often very narrow. It might be drought or rain, hail or sunshine, grasshoppers or absence of insects, and health or disease among livestock that made the difference between bankruptcy and survival."[10]

Luck played a role in separating those who stayed from those who left. For some it rained but didn't hail, the grasshoppers didn't descend, and they got good yields in the years when prices were high. Inertia probably meant some-

North Dakota farmer plowing with a four-horse team around 1900. State Historical Society of North Dakota (00090-011). Used by permission.

thing as well; it takes more energy to move than it does to remain stationary. Plains people who stayed frequently congratulated themselves for their superior character. Reflecting on mobility and persistence in the Nebraska Sand Hills, Mari Sandoz observed that "the shiftless, and those who live from the prosperity of their fellows, drifted to greener fields early. Only the strong and courageous, the ingenious and the stubborn, remained." There was probably something to that, but beyond luck and character, which are difficult to measure, other factors stood out.[11]

Immigrants were usually more likely than older-stock Americans to stay put. Harger thought that Mennonites in Kansas persisted because they were better farmers than their neighbors. Whether or not that was the case, immigrants were more likely than natives to come with supportive neighbors and kin. Living in an alien country with practices and laws with which they were unfamiliar, it was riskier for them to move on; living far from their places of birth or alienated from those places, it was harder for them to go back. One result of immigrant persistence was that communities with an immigrant presence sometimes became immigrant communities, as nonimmigrants sold out to immigrant neighbors or to new arrivals practicing chain migration. In the Four Corners community in Kansas, older-stock Americans moved on, selling their property to German neighbors, with the result that, within a few years, what had been a mixed community became a German one. Among native-born and immigrant settlers alike, persisters were more likely than nonpersisters to be older, be married, and own substantial property. Such people had a more tangible stake in the community, they were presumably more prudent and conservative, and the price of failure for them was higher in terms of the reproach they felt from their families and the cost of starting over.[12]

The high level of physical mobility on the Great Plains may have impeded the formation and maintenance of community. In Boone County, Nebraska, rapid population turnover appeared to deprive women of "the comfort of trusted social networks," and everywhere the "restless movement toward new opportunity tended to dissolve neighborhood and social ties as quickly as it formed them." The materialism and commercial motivations of plains settlers also may have retarded community formation: "The strength of the money-motive in the early agriculture of the Plains was a main factor holding back the development of strong community life . . . on the whole, the settlers were a race of land-hungry individualists."[13]

It is logical that mobility threatened community, but community formation and maintenance is a complicated issue. Americans were a mobile people generally, and the rich tradition of rural neighborliness endured and even thrived in the face of that reality. Indeed, the rapid formation of neighborly connections might have been more necessary in a highly mobile community than in a more stable one. Moreover, most newcomers found a ready-made community with schools, churches, fraternal organizations, and other institutions just like those in the places they had left—a benefit of the tendency of frontier neighborhoods to reproduce the culture of home. It is also true that a strong community can exist without including every person or family living in a particular area. Indeed, community is often as much about who is excluded as it is about who is included. High population turnover does not prove that the sense of community among more established residents was weak or absent.

The notion that materialism and the desire for commercial success imperiled community formation and maintenance is even more problematic. Most Americans, including those in newly settled areas, were materialistic and market-oriented, a reality on which immigrants commented, often disapprovingly. But did their economic values and behavior prevent them from forming neighborhood social networks, exchanging work, helping one another in times of trouble, and uniting to create churches, schools, cooperatives, and other social institutions? There is little evidence that it did. Farmers did not see themselves as competing with one another. They usually believed that, as a group, they were competing with the railroads, elevators, and bankers who were taking money from them through manipulation and price gouging. Moreover, in common with other human beings, farmers had the ability to hold logically incompatible ideas in their minds simultaneously and to act on them. Thus, while a materialistic regard for self and an altruistic regard for community were logically inconsistent, as a practical matter plains farmers could, and did, exhibit both. Indeed, could they have built schools and churches, supported clubs and lodges, created co-ops, and maintained their communities in a variety of other ways had they not achieved at least some commercial success?[14]

Maturity on the Eastern Plains

Population inflows and outflows remained a reality on the eastern plains, as was the case in most of the rest of the country, but by 1900 the region had achieved a level of stability it had not known previously. Improved economic performance was the key. To a significant degree, the farm economy throughout the plains benefited from rising prices for agricultural commodities after 1900. But the parts of the region that had been settled during the boom of the 1880s were now better able to take advantage of the opportunities the market presented.

Farmers had "mastered the art of agriculture" on the plains. To some extent the adjustments they made came through a process of individual and neighborhood experimentation involving trial and error. Farmers also benefited from self-taught agricultural experts, such as Hardy Webster Campbell, whose "dry farming" system involved modifications in planting and cultivation designed to help farmers make crops in seasons and regions with low availability of moisture.[15]

More effective farming did not depend solely on the efforts of farmers and amateur experimenters. The railroads, whose future prosperity and even survival required a prosperous and productive agriculture in the region, sponsored scientific experimentation, bankrolled efforts to address regional problems (such as the pro-irrigation Dry Farming Congress), ran "demonstration trains" to acquaint farmers with new developments in plant production and animal breeding, paid premiums to farmers willing to use new methods, and even distributed purebred bulls to upgrade local cattle herds. The railroads were self-interested, to be sure, but their self-interest was sometimes beneficial to farmers.[16]

Federal, state, and local governments also played a role in improving farming on the Great Plains. Indeed, one of the ironies of the plains agricultural experience was that, their self-conscious independence and individualism notwithstanding, plains settlers enjoyed more government aid than had the residents of any previous American frontier. The U.S. Department of Agriculture (USDA), created in that pivotal year of 1862, addressed numerous problems of plains agriculture, especially through its bureaus of Plant Industry and Animal Industry. Both could boast of noteworthy accomplishments. The former scoured the globe for appropriate crop varieties for the plains environment and introduced such new crops as durum wheat, a drought-resistant variety especially well suited for making pasta. The Bureau of Animal Industry discovered the cause of "Texas Fever," which sometimes devastated cattle herds on the plains, and devised a solution for it.

For their part, the states inspected seed and certified it for purity and germination, tested agricultural chemicals and fertilizers for consistency and potency, quarantined diseased animals, and maintained agricultural statistics. The state and federal governments cooperated in providing support for agricultural experiment stations, which played a critical role in improving farming and ranching on the Great Plains. Experiment stations were created by the Hatch Act of 1887. Congress attached them to existing land-grant colleges and gave them the mission of addressing the particular agricultural challenges confronting the states in which they were located. Early experiment stations left much to be desired from the point of view of scientific expertise and rigor, but they tested new crops and new varieties of old crops, undertook selective breeding of livestock, attempted to improve farmers' business practices, and developed experimental programs in such areas as soil chemistry, irrigation, veterinary medicine, animal nutrition, and entomology. Their experimental work was hit or miss, but the results generally benefited farmers on the plains. To carry the results of experimentation to farmers, states published agricultural bulletins that were distributed at no cost to farmers and sponsored "farmers' institutes." These were local meetings, usually held after the growing season, at which experiment station scientists shared their latest findings with interested farmers. Taken together, these government efforts vindicated the Republican argument, embodied in the legislative program of 1862, that government aid for individuals helped them prosper while enriching the entire country.[17]

All of these public and private efforts helped make agriculture more productive and predictable, though the Great Plains drought of the 1930s made it clear that Charles Moreau Harger was overstating the case in 1904 when he proclaimed that farmers in the region faced "little probability of failure." Agriculture on the plains was now also more stable, though perfect stability could never be achieved in a modern, dynamic, capitalist system. Indeed, even as observers were touting the maturity of the plains and its agriculture, implement manufacturers and agricultural engineers were preparing to introduce machinery that would eventually revolutionize the region. Still, stability and maturity characterized agriculture on the plains in the early twentieth century to a much greater extent than it had during the early years of settlement.[18]

Farmers' economic positions improved as they learned from government, the railroads, and neighbors. Another key factor in their brightening prospects was their steady addition of more acres to crop production. Breaking land was a slow process, and homesteaders had seldom broken more than forty to fifty acres when they proved up. Over the years they brought more of their acres into crop production, especially when railroad transportation and handling and marketing facilities became more convenient. Sometimes cultivation increased dramatically. In Smith County, in north-central Kansas, the average farm had only 57 improved acres in 1880; ten years later it had 141. With more land in cultivation, farmers could devote more resources to the production of cash crops that flowed into international commodity markets. When prices were good, larger acreages translated into higher incomes. Farms with more land in crop production also demanded more of the owner's attention, limiting the time he or she could devote to off-farm employment. Nineteenth- and early-twentieth-century farming remained a risky business because there were no government programs to insure farmers against crop failures or buffer them from market fluctuations. To diminish their risk, farm families continued to strive to maximize their self-sufficiency and diversify their income flows. Most families maintained large gardens and potato patches and kept livestock for home consumption, and women continued to make their own soap and to produce eggs, butter, and cream for local markets. Sometimes these barnyard products saved farms from economic catastrophe.[19]

Commercial agriculture on the plains revolved mainly around cattle and wheat. Cattle had been a major factor in plains life from the beginning of settlement, when ranchers replaced wild, grass-eating ruminant animals—bison—with domesticated, grass-eating ruminant animals. Cattle continued to be important, especially on the western plains, because much native prairie remained unbroken and because cattle and grain prices usually moved in opposite directions, buffering diversified producers against catastrophic losses.[20]

By 1900 most of the eastern and central plains constituted the Wheat Belt. Wheat had much to recommend it. Wheat is a grass that grows well in a natural grassland. In a semi-arid climate, which characterizes much of the plains, wheat is less vulnerable to rust and other fungal outbreaks than it is in more humid areas, and winter wheat, planted in the fall and harvested in June or July, makes maximum use of available moisture. The attractiveness of the crop was enhanced by the introduction of new, Eurasian varieties, such as durum and Turkey Red, which yielded well in dry conditions. In contrast to corn, a significant alternative crop in some areas, wheat required little labor except at harvest and lent itself to mechanized production. Finally, there was strong national and international demand for wheat, and a well-established handling, marketing, and transportation infrastructure was dedicated to it. By 1910 Kansas and North Dakota were the leading wheat-producing states in the country.[21]

Sarah Comstock, Charles Moreau Harger, and others may have exaggerated the comfort and grace of life on the plains after 1900, but greater prosperity did translate into improved living standards. Farm families built more comfortable and modern homes, often purchased from mail-order houses and shipped by rail to be assembled on site. They furnished their homes with fashionable furniture, carpets, and wall hangings. They had the discretionary income to buy consumer products that made their lives brighter and easier. Whether they took advantage of the opportunities prosperity afforded depended on the family. Visiting the northern plains in 1908, A. E. Dickey wrote approvingly of the Garrett family, which had a substantial house, a telephone, a gramophone, and a second-hand piano. For their neighbors, the prosperous Hackers, however, there was nothing "except incessant toil day after day" in a spare and stultifying environment. Old habits died hard on the plains.[22]

Observers especially emphasized enhancements in the lives of women on the Great Plains, who a few years before had been broadly portrayed as victims of unremitting toil and isolation. The Country Life movement, which developed after the turn of the century, pressed for the improvement of all rural women's lives in a variety of ways. Home economists from the land-grant colleges and the USDA emphasized making houses more convenient for farm women and beautifying both indoor and outdoor environments. Ladies' and farm magazines and farmers' institutes' winter programs also provided suggestions for improvements. Whether they were responding to these prompts or simply acting on their own inclinations, many plainswomen took advantage of prosperity to beautify their homes and yards and adopt conveniences such as sewing machines, mechanical cream separators, and even gasoline-powered washing machines. Although these changes commanded a good deal of attention in national publications, rural women lagged far behind their urban sisters when it came to the ease, convenience, and grace of living. The lyrical descriptions of plainswomen's material lives by visiting journalists were belied by the reality that many women continued to labor in surroundings that city people considered primitive, making soap, canning in steaming kitchens for days, cleaning stalls and chicken houses, cooking and baking for large harvest crews—all usually without electricity or even water in the house.[23]

People living in long-settled areas of the plains also took advantage of their relative prosperity to enhance their institutions. They remodeled their churches and upgraded their schools, often lengthening school terms and hiring better-trained teachers. They formed and joined more lodges and clubs. And they improved their roads, facilitating travel to town, which they sometimes now undertook in automobiles.

Life on the plains did not approach the grace and convenience of urban living in the early twentieth century, nor did rural institutions offer what urban institutions did. But veteran residents of the plains could justly take pride in the progress their communities had made since the early sod busting days.

The Passing of the Settler Generation

Ironically, the improvements in agriculture, living standards, and institutional development that had marked the maturation of the eastern and central plains carried the seeds of instability in the form of continuing population turnover. While physical mobility here did not match the levels of the early settlement years or of the newly settled western plains, it was pronounced enough to elicit widespread comment. Some of those moving out were selling their farms at a profit so that they could relocate farther west, where their dollars could command more—and, alas, less dependable—acres. Federal lands were more plentiful on the western plains, and many ranchers were taking advantage of the agricultural boom to subdivide lands for sale to farmers. Other large landholders, such as land companies and "bonanza" farms—defined as single units of at least three thousand acres—were also breaking up their holdings for sale to smaller farmers.

The life cycle further contributed to farm turnover. In addition to those who were moving out and moving on, numerous farmers on the eastern and central plains were retiring and living on the proceeds from the sale or rental of their farms. Farmers who had taken land during the boom of the early eighties were by 1910 reaching the age at which farming became too strenuous. Some sold to one or more of their children, holding a mortgage to protect other heirs—who would thereby be guaranteed their share of the farm in the event of the parents' death—and living on the mortgage payments. Others sold to a nonfamily member or rented out the land and lived off the rental income. Those who did so could take satisfaction in the fruits of their labor. Their efforts and sacrifices had raised the value of their land, but that had not been the only factor. The agricultural boom had helped, as had the development of marketing and transportation facilities. Just having a railroad in a county raised the price of land by nearly 25 percent. For all of these reasons, land values rose impressively. In three southeastern Nebraska counties, the value of the average farm rose from $1,664 in 1880, early in the first boom, to $16,741 in 1910, an increase of over 1,000 percent! It is often said that the farmer's only real payday is the day he sells his farm. For the members of the settler generation who were still around, 1910 was a good year for a payday.[24]

Epilogue

THE PROCESS of plains settlement is an American success story. Settlers came to a strange and challenging region from areas to the east and from Europe. They struggled to find wood and water. They created farms out of a difficult and sometimes unyielding land. They worked to maintain shelter and to feed and clothe themselves and their families. They spent years making farms that could support them and eventually provide them with the comforts and small luxuries of life. They built institutions that enriched their lives emotionally and intellectually. They had help from government, from families, from friends and neighbors, and sometimes from railroads and other businesses, but they succeeded mainly because of their own hard work and determination. As they reflected on their lives, they could take pride in their success, not least because so many of their fellow settlers on the Great Plains had failed.

Throughout the plains there are "century farms" that have been in the same family for at least one hundred years. The owners of these farms take justifiable pride in their forebears, people of courage and determination who came to a difficult place and created a legacy for future generations. They honor their ancestors in county histories and local historical societies and by struggling to maintain the institutions—especially the churches—the settler generation created. Their families are enduring embodiments of the success of the settlement endeavor on the Great Plains. But success took other forms as well.

By 1910 Rachel and Abraham Calof lived in a frame house, lighted by kerosene and heated with coal, and were well and abundantly fed. They had expanded their farm, and Abraham had made it profitable through good management and his skill as a horse breaker. They were leaders in the local Jewish community. Abraham was instrumental in getting the first school built and served on the school board. Yet in 1917 they sold their farm and moved to Saint Paul, where Abraham used the profits of the family's hard and often heroic labor to open a grocery store. Theirs did not turn out to be a century farm, but their experience, too, is testimony to the ultimate success of the settlement endeavor on the Great Plains.

NOTES

PROLOGUE

1. Anne Kelsch, "Alexander Henry and the Historical Landscape," www.angelfire.com/bug/rud/KELSCH.htm.

2. J. Sanford Rikoon, ed., *Rachel Calof's Story: Jewish Homesteader on the Northern Plains* (Bloomington: Indiana University Press, 1995), 33, 39–40, 41, 68, 76.

3. Mary Dodge Woodward, *The Checkered Years*, ed. Mary Boynton Cowdrey (Cass County Historical Society, no place or date); Roderick Cameron, *Pioneer Days in Kansas: A Homesteader's Narrative of Early Settlement and Farm Development on the High Plains Country of Northwest Kansas* (Belleville, KS: Cameron Book Company, 1951).

4. Kenneth Wiggins Porter, ed., "'Holding Down' a Northwest Kansas Claim, 1885–1888," *Kansas Historical Quarterly* 22 (1956): 220–235; Mrs. Raymond Millbrook, ed., "Mrs. Hattie E. Lee's Story of Her Life in Western Kansas," *Kansas Historical Quarterly* 22 (1956): 114–137; Jane Taylor Nelsen, ed., *A Prairie Populist: The Memoirs of Luna Kellie* (Iowa City: University of Iowa Press, 1992); Seth K. Humphrey, *Following the Prairie Frontier* (Minneapolis: University of Minnesota Press, 1931).

5. Elliott West, *The Contested Plains: Indians, Goldseekers, and the Rush to Colorado* (Lawrence: University Press of Kansas, 2000).

6. G. M. Whicher, "Phrases of Western Life: II—Farm Life in Western Nebraska," *Outlook* 49 (13 January 1894): 63; E. V. Smalley, "The Isolation of Life on Prairie Farms," *Atlantic Monthly* 72 (September 1893): 379; Laura Ingalls Wilder, *The Long Winter* (New York: Harper Collins, 1953), 20; Seth K. Humphrey, *Following the Prairie Frontier* (Minneapolis: University of Minnesota Press, 1931), 114.

7. Elwyn B. Robinson, *History of North Dakota* (Lincoln: University of Nebraska Press, 1966), 8.

8. Heather Cox Richardson, *The Greatest Nation of the Earth: Republican Economic Policies during the Civil War* (Cambridge: Harvard University Press, 1997), 2.

9. Stephen J. Rockwell, *Indian Affairs and the Administrative State in the Nineteenth Century* (Cambridge: Cambridge University Press, 2010), 1.

10. Bruce Garver, "Immigration to the Great Plains, 1865–1914: War, Politics, Technology, and Economic Development," *Great Plains Quarterly* 31 (Summer 2011): 182.

11. Eric Rauchway, *Blessed among Nations: How the World Made America* (New York: Hill and Wang, 2007), 58.

CHAPTER ONE: How They Acquired Land

1. Paul Wallace Gates, *Fifty Million Acres: Conflicts over Kansas Land Policy, 1854–1890* (New York: Atherton Press, 1966), 3.

2. Stephen J. Rockwell, *Indian Affairs and the Administrative State in the Nineteenth Century* (Cambridge: Cambridge University Press, 2010), 296–297; Nathan B. Sanderson, "'We Were All Trespassers': George Edward Lemmon, Anglo-American Cattle Ranching, and the Great Sioux Reservation," *Agricultural History* 85 (Winter 2011): 50–71.

3. My understanding of federal land policy draws heavily on three classics, Paul W. Gates, *History of Public Land Law Development* (Washington: U.S. Government Printing Office, 1968); Benjamin Horace Hibbard, *A History of Public Land Law Policies* (New York: Macmillan, 1924); and Roy M. Robbins, *Our Landed Heritage: The Public Domain, 1776–1936* (Princeton: Princeton University Press, 1942).

4. Heather Cox Richardson, *The Greatest Nation of the Earth: Republican Economic Policies during the Civil War* (Cambridge: Harvard University Press, 1997), 143.

5. Seth K. Humphrey, *Following the Prairie Frontier* (Minneapolis: University of Minnesota Press, 1931), 80.

6. Willard W. Cochrane, *The Development of American Agriculture: A Historical Analysis* (Minneapolis: University of Minnesota Press, 1979), 83; James Fredric Hamburg, *The Influence of Railroads upon the Processes and Patterns of Settlement in South Dakota* (New York: Arno Press, 1981), 239.

7. Mari Sandoz, *Old Jules* (Boston: Little, Brown, 1935), 270. For the functions of locators see Mary W. M. Hargreaves, *Dry Farming in the Northern Great Plains, 1900–1925* (Cambridge: Harvard University Press, 1957), 413–414.

8. Mabel Lewis Stuart, "The Lady Honyocker: How Girls Take Up Claims and Make Their Own Homes on the Prairie, " *Independent* 75 (17 July 1913): 135–136; Lori Ann Lahlum, "Mina Westbye: Norwegian Immigrant, North Dakota Homesteader, Studio Photographer, 'New Women,'" *Montana: The Magazine of Western History* 60 (Winter 2010): 5; J. Sanford Rikoon, ed., *Rachel Calof's Story: Jewish Homesteader on the Northern Plains* (Bloomington: Indiana University Press, 1995), 35; Sherry L. Smith, "Single Women Homesteaders: The Perplexing Case of Elinore Pruitt Stewart," in Sandra K. Schackel, ed., *Western Women's Lives: Continuity and Change in the Twentieth Century* (Albuquerque: University of New Mexico Press, 2003), 161–182. For a thorough study of the female homesteading experience in one state, see H. Elaine Lindgren, *Land in Her Own Name: Women as Homesteaders in North Dakota* (Fargo: North Dakota Institute for Regional Studies, 1991).

9. Hamlin Garland, in *A Son of the Middle Border* (New York: Macmillan, 1923), 303, notes that he marked his land with a straddle-bug in his short-lived homesteading adventure.

10. James R. Beck, "Homesteading in Union Township, Clay County, Kansas, 1863–1889," *Kansas History* 34 (Autumn 2011): 197, 191; Sandoz, *Old Jules*, 85. For challenges to homesteaders see Lindgren, *Land in Her Own Name*, 78–79. For land office and proving-up difficulties, see Craig Miner, *West of Wichita: Settling the High Plains of Kansas, 1865–1890* (Lawrence: University Press of Kansas, 1986), 138.

11. Beck, "Homesteading in Union Township," 193.

12. Lahlum, "Mina Westbye," 11, 14; Lloyd A. Svendsbye, *I Paid All My Debts . . . : A Norwegian-American Immigrant Saga of Life of the Prairie of North Dakota* (Minneapolis: Lutheran University Press, 2009), 39–40.

13. Humphrey, *Following the Prairie Frontier*, 132, 80; Gates, *Fifty Million Acres*, 244.

14. Paula M. Nelson, *After the West Was Won: Homesteaders and Town-Builders in Western South Dakota, 1900–1917* (Iowa City: University of Iowa Press, 1986), 42; Lahlum, "Mina Westbye," 14; Katherine Harris, *Long Vistas: Women and Families on Colorado Homesteads* (Niwot, CO: University Press of Colorado, 1993), 149; Mary Isabel Brush, "Women on the Prairies," *Collier's* 46 (28 January 1911): 16; Lindgren, *Land in Her Own Name*, 67.

15. R. Douglas Hurt, *The Big Empty: The Great Plains in the Twentieth Century* (Tucson: University of Arizona Press, 2011), 21.

16. Fred A. Shannon, *The Farmer's Last Frontier: Agriculture, 1860–1897* (New York: Rinehart, 1945), 59. For more on the Timber Culture Act, see Gates, *Public Land Law Development*, 399–400, and Hibbard, *History of Public Land Policies*, 418.

17. Roderick Cameron, *Pioneer Days in Kansas: A Homesteader's Narrative of Early Settlement and Farm Development on the High Plains Country of Northwest Kansas* (Belleville, KS: Cameron Book, 1951), 14–15; Gilbert C. Fite, *The Farmer's Frontier, 1865–1890* (New York: Holt, Rinehart and Winston, 1966), 22.

18. Robert V. Hine, *The American West: An Interpretive History* (Boston: Little, Brown, 1984), 177; Cochrane, *The Development of American Agriculture*, 83.

19. Christopher S. Decker and David T. Flynn, "The Railroad's Impact on Land Values in the Upper Great Plains at the Closing of the Frontier," *Historical Methods* 40 (Winter 2007): 33; Fite, *The Farmer's Frontier*, 18.

20. Gates, *Public Land Law Development*, 282, 335; Hibbard, *History of Public Land Policies*, 332; Robbins, *Our Landed Heritage*, 246–47; and Shannon, *The Farmer's Last Frontier*, 56–68.

21. Shannon, *The Farmer's Last Frontier*, 64–65.

22. For railroad development programs generally, see Roy V. Scott, *Railroad Development Programs in the Twentieth Century* (Ames: Iowa State University Press, 1985).

23. Miner, *West of Wichita*, 199; Hargreaves, *Dry Farming*, 403; Gates, *Fifty Million Acres*, 286–287; Gates, *Public Land Law Development*, 213.

24. Arthur M. Johnson and Barry E. Supple, *Boston Capitalists and Western Railroads: A Study in the Nineteenth-Century Railroad Investment Process* (Cambridge: Harvard University Press, 1967), 295; James B. Hedges, "The Colonizing Work of the

Northern Pacific Railroad," *Mississippi Valley Historical Review* 13 (1926): 330; David M. Emmons, *Garden in the Grasslands: Boomer Literature in the Central Great Plains* (Lincoln: University of Nebraska Press, 1971), 88–89; Ian Frazier, *Great Plains* (New York: Farrar/Straus/Giroux, 1989), 168, 191.

25. Fite, *The Farmers' Frontier*, 31; Richard C. Overton, *Burlington West: A Colonization History of the Burlington Railroad* (Cambridge: Harvard University Press, 1941), 336; Hedges, "The Northern Pacific Railroad," 321; Scott, *Railroad Development Programs*, 30.

26. Gates, *Fifty Million Acres, 273;* Baldwin F. Kruse, *Paradise on the Prairie: Nebraska Settler Stories* (Lincoln, NE: Paradise Publishing, 1986), 37; Allan G. Bogue, *Money at Interest: The Farm Mortgage on the Middle Border* (Lincoln: University of Nebraska Press, 1955), 214; Shannon, *The Farmers' Last Frontier*, 42; Overton, *Burlington West*, 314–315; Hedges, "The Northern Pacific Railroad," 312.

27. John Radzilowski, "A New Poland in the Old Northwest: Polish Farming Communities on the Northern Great Plains, " *Polish American Studies* 59 (Autumn 2002): 81–82.

28. Spectator, "Home Seekers' Excursion," *Outlook* 86 (1 June 1907): 235.

29. Johnson and Supple, *Boston Capitalists and Western Railroads*, 296; Hiram M. Drache, *The Challenge of the Prairie: Life and Times of Red River Pioneers* (Fargo: North Dakota Institute for Regional Studies, 1970), 18. See also Miner, *West of Wichita*, 68.

CHAPTER TWO: How They Built Farms

1. J. Sanford Rikoon, ed., *Rachel Calof's Story: Jewish Homesteader on the Northern Plains* (Bloomington: Indiana University Press, 1995), 22–77.

2. Spectator, "Home Seekers' Excursion," *Outlook* 86 (1 June 1907): 236; Seth K. Humphrey, *Following the Prairie Frontier* (Minneapolis: University of Minnesota Press, 1931), 132.

3. Nina Farley Wishek, *Along the Trails of Yesterday: A History of McIntosh County* (Ashley, ND: Ashley Tribune, 1941), 98.

4. Joanna L. Stratton, *Pioneer Women: Voices from the Kansas Frontier* (New York: Simon and Schuster, 1981), 53; Mrs. Raymond Millbrook, ed., "Mrs. Hattie E. Lee's Story of Her Life in Western Kansas," *Kansas Historical Quarterly* 22 (1956): 116, 119.

5. Katherine Harris, *Long Vistas: Women and Families on Colorado Homesteads* (Niwot, CO: University Press of Colorado, 1993), 93–97.

6. Everett Dick, "Sunbonnet and Calico: The Homesteader's Consort," *Nebraska History* 10 (March 1966): 7, 5; Walker D. Wyman, *Frontier Woman: The Life of a Woman Homesteader on the Dakota Frontier* (River Falls, WI: University of Wisconsin—River Falls Press, 1972), 16; Baldwin F. Kruse, *Paradise on the Prairie: Nebraska Settler Stories* (Lincoln, NE: Paradise Publishing, 1986), 29–30; I. E. M. Smith, "Two City Girls' Experiences in Holding Down a Claim: A Montana Pastoral," *Overland Monthly* 24 (August 1894): 147, 148, 151.

7. Roderick Cameron, *Pioneer Days in Kansas: A Homesteader's Narrative of Early*

Settlement and Farm Development on the High Plains Country of Northwest Kansas (Belleville, KS: Cameron Book, 1951), 16, 19.

8. Fred W. Peterson, *Homes in the Heartland: Balloon Frame Farmhouses of the Upper Midwest, 1850–1920* (Lawrence: University Press of Kansas, 1992), 55–56; Paula M. Nelson, *After the West Was Won: Homesteaders and Town-Builders in Western South Dakota, 1900–1917* (Iowa City: University of Iowa Press, 1986), 28, 30; Linda Peavy and Ursula Smith, *Pioneer Women: The Lives of Women on the Frontier* (Norman: University of Oklahoma Press, 1996), 51, 60.

9. Hamlin Garland, "A Prairie Heroine," in Donald Pizer, ed., *Hamlin Garland, Prairie Radical: Writings from the 1890s* (Urbana: University of Illinois Press, 2010), 36.

10. Elizabeth Hampsten, *Read This Only to Yourself: The Private Writings of Midwestern Women, 1890–1910* (Bloomington: Indiana University Press, 1982), 193; Charles Moreau Harger, "Phases of Western Life: III.-Winter on the Prairies," *Outlook* 49 (20 January 1894): 126; Steven Kinsella, *900 Miles from Nowhere: Voices from the Homestead Frontier* (St. Paul: Minnesota Historical Society Press, 2006), 77.

11. Kenneth Wiggins Porter, ed., "'Holding Down' a Northwest Kansas Claim, 1885–1888," *Kansas Historical Quarterly* 22 (1956): 225; Harris, *Long Vistas*, 107–108; Dick, "Sunbonnet and Calico," 6; Humphrey, *Following the Prairie Frontier*, 129.

12. Cameron, *Pioneer Days in Kansas*, 31; Peavy and Smith, *Pioneer Women*, 57.

13. G. M. Whicher, "Phases of Western Life: II. Farm Life in Western Nebraska," *Outlook* 49 (13 January 1894): 64; Mary W. M. Hargreaves, *Dry Farming in the Northern Great Plains, 1900–1925* (Cambridge: Harvard University Press, 1957), 61, 500; Jane Taylor Nelsen, ed., *A Prairie Populist: The Memoir of Luna Kellie* (Iowa City: University of Iowa Press, 1992), 30.

14. Kruse, *Paradise on the Prairie*, 40; Porter, "A Northwest Kansas Claim," 230; L. L. Bloomenshine, *Prairie around Me* (San Diego: Raphael Publications, 1972), 2.

15. Harris, *Long Vistas*, 106.

16. David Laskin, *The Children's Blizzard* (New York: Harper, 2005); Humphrey, *Following the Prairie Frontier*, 82.

17. Venola Lewis Bivans, ed., "The Diary of Luna E. Warner, a Kansas Teenager of the Early 1870s," *Kansas Historical Quarterly* 35 (1969): 281; Wishek, *Along the Trails of Yesterday*, 206.

18. John D. McDermott, "The Plains Forts: A Harsh Environment," *Nebraska History* 91 (Spring 2010): 11.

19. June Granatir Alexander, *Daily Life in Immigrant America, 1870–1920: How the Second Great Wave of Immigrants Made Their Way in America* (Chicago: Ivan R. Dee, 2009), 74; Stratton, *Pioneer Women*, 103, 106; Emily Greene Balch, "The True Story of a Bohemian Pioneer," *Chautauquan* 49 (February 1908): 401; Nelsen, *A Prairie Populist*, 22–23.

20. Craig Miner, *Next Year Country: Dust to Dust in Western Kansas, 1890–1940* (Lawrence: University Press of Kansas, 2006), 61.

21. Hamlin Garland, *A Son of the Middle Border* (New York: MacMillan, 1923),

132; O. E. Rölvaag, *Giants in the Earth: A Saga of the Prairies* (New York: Harper and Row, 1927), 101.

22. Humphrey, *Following the Prairie Frontier*, 131.

23. Stratton, *Pioneer Women*, 86; Mary Dodge Woodward, *The Checkered Years*, ed. Mary Boynton Cowdrey (Cass County Historical Society, n.d.), 120; Mari Sandoz, *Old Jules* (Boston: Little, Brown, 1935), 82–83; Rikoon, *Rachel Calof's Story*, 64–65.

24. Julie Roy Jeffrey, *Frontier Women: The Trans-Mississippi West, 1840–1880* (New York: Hill and Wang, 1979), 10; Rikoon, *Rachel Calof's Story*, 45, 60; Andrea G. Radke, "Refining Rural Spaces: Women and Vernacular Gentility in the Great Plains, 1880–1920," *Great Plains Quarterly* 24 (Fall 2004): 230; Faye C. Lewis, *Nothing to Make a Shadow* (Ames: Iowa State University Press, 1971), 72.

25. Nelson, *After the West Was Won*, 51; Robert C. Ostergren, *A Community Transplanted: The Trans-Atlantic Experience of a Swedish Immigrant Settlement in the Upper Middle West, 1835–1915* (Madison: University of Wisconsin Press, 1988), 189.

26. Cameron, *Pioneer Days in Kansas*, 54.

27. Hiram M. Drache, *The Challenge of the Prairie: Life and Times of Red River Pioneers* (Fargo: North Dakota Institute for Regional Studies, 1970), 60–61.

28. Homer E. Socolofsky, "Kansas in 1876," *Kansas Historical Quarterly* 43 (Spring 1977): 6; Lewis, *Nothing to Make a Shadow*, 39; Geoff Cunfer, *On the Great Plains: Agriculture and Environment* (College Station: Texas A. and M. University Press, 2005), 17: Richard C. Overton, *Burlington West: A Colonization History of the Burlington Railroad* (Cambridge: Harvard University Press, 1941), 338; Kinsella, *900 Miles from Nowhere*, 82; Drache, *Challenge of the Prairie*, 61.

29. Socolofsky, "Kansas in 1876," 28; Joshua D. McFadyen, "Breaking Sod or Breaking Even? Flax on the Northern Great Plains and Prairies, 1889–1930," *Agricultural History* 83 (Spring 2009): 221–246; David B. Danbom, *"Our Purpose Is to Serve": The First Century of the North Dakota Agricultural Experiment Station* (Fargo: North Dakota Institute for Regional Studies, 1990), 37.

30. Cunfer, *On the Great Plains*, 8, 112–115. See also Kenneth M. Sylvester, "Ecological Frontiers on the Grasslands of Kansas: Changes in Farm Scale and Crop Diversity," *Journal of Economic History* 69 (December 2009): 1041–1062.

31. Walter Nugent, *Into the West: The Story of Its People* (New York: Alfred A. Knopf, 1999), 144; David B. Danbom, *The Resisted Revolution: Urban American and the Industrialization of Agriculture, 1900–1930* (Ames: Iowa State University Press, 1979): 37–38.

32. Bloomenshine, *Prairie around Me*, 117.

33. Wishek, *Along the Trails of Yesterday*, 235.

34. Drache, *Challenge of the Prairie*, 69, 186–187; Arthur M. Johnson and Barry E. Supple, *Boston Capitalists and Western Railroads: A Study in the Nineteenth Century Railroad Investment Process* (Cambridge: Harvard University Press, 1967), 297; Gilbert C. Fite, *The Farmer's Frontier, 1865–1900* (New York: Holt, Rinehart and Winston, 1966), 43; Robert Nesbit, *History of Wisconsin*, vol. 3, *Urbanization and Industrialization, 1873–1893* (Madison: Wisconsin Historical Press, 1985), 541.

35. Nelsen, *A Prairie Populist*, 55; Sandoz, *Old Jules*, 89–90. For butter and egg money and its contributions to the settlers' success, see Barbara Handy-Marchello, *Women of the Northern Plains: Gender and Settlement on the Homestead Frontier, 1870–1930* (St. Paul: Minnesota Historical Society Press, 2005).

36. Rikoon, *Rachel Calof's Story*, 30–31; Nelsen, *A Prairie Populist*, 40, 45; H. Elaine Lindgren, *Land in Her Own Name: Women as Homesteaders in North Dakota* (Fargo: North Dakota Institute for Regional Studies, 1991), 93; Dorothy Hubbard Schwieder, *Growing Up with the Town: Family and Community on the Great Plains* (Iowa City: University of Iowa Press, 2002), 15; Harris, *Long Vistas*, 137.

37. Scott G. and Sally Allen McNall, *Plains Families: Exploring Sociology through Social History* (New York: St. Martin's Press, 1983), 44; Lloyd A. Svendsbye, *I Paid All My Debts . . . : A Norwegian-American Saga of Life on the Prairie of North Dakota* (Minneapolis: Lutheran University Press, 2009), 41; Cameron, *Pioneer Days in Kansas*, 116.

38. Nelson, *After the West Was Won*, 42–43; Lindgren, *Land in Her Own Name*, 12; Lori An Lahlum, "Mina Westbye: Norwegian Immigrant, North Dakota Homesteader, Studio Photographer, and 'New Woman,'" *Montana: The Magazine of Western History* 60 (Winter 2010): 8; Schwieder, *Growing Up with the Town*, 15; and Woodward, *The Checkered Years*, 182.

39. Willa Gibert Cather, *My Antonia* (Lincoln: University of Nebraska Press, 1994), 191; Mary Hurlbut Cordier, *Schoolwomen of the Prairies and Plains: Personal Narratives from Iowa, Kansas, and Nebraska, 1860s–1920s* (Albuquerque: University of New Mexico Press, 1992), 21; Porter, "A Northwest Kansas Claim," 225; Millbrook, "Mrs. Hattie E. Lee's Story," 116, 122–123, 126.

CHAPTER THREE: How They Got Credit

1. Lloyd A. Svendsbye, *I Paid All My Debts . . . : A Norwegian-American Immigrant Saga of Life on the Prairie of North Dakota* (Minneapolis: Lutheran University Press, 2009), 39–40; Jane Taylor Nelsen, ed., *A Prairie Populist: The Memoirs of Luna Kellie* (Iowa City, IA: University of Iowa Press, 1992), 121.

2. Nelsen, *A Prairie Populist*, 121.

3. June Granatir Alexander, *Daily Life in Immigrant America, 1870–1920: How the Second Great Wave of Immigrants Made Their Way in America* (Chicago: Ivan R. Dee, 2009), 63; Hiram M. Drache, *The Challenge of the Prairie: Life and Times of Red River Pioneers* (Fargo: North Dakota Institute for Regional Studies, 1970), 191; Roderick Cameron, *Pioneer Days in Kansas: A Homesteader's Narrative of Early Settlement and Farm Development on the High Plains Country of Northwest Kansas* (Belleville, KS: Cameron Book, 1951), 99.

4. Eleanor H. Hinman and J. O. Rankin, *Farm Mortgage History of Eleven Southeastern Nebraska Townships, 1870–1932* (Lincoln: College of Agriculture, University of Nebraska, Agricultural Experiment Station Research Bulletin 67, August, 1933), 67.

5. Svendsbye, *I Paid All My Debts*, 37, 40.

6. Fred A. Shannon, *The Farmers' Last Frontier: Agriculture, 1860–1897* (New York: Rinehart, 1945), 188.

7. Allan G. Bogue, Brian Q. Cannon, and Kenneth J. Winkle Jr., "Oxen to Organs: Chattel Credit in Springdale Town, 1849–1900," *Agricultural History* 77 (Summer 2003): 420–452; Stanley B. Parsons, *The Populist Context: Rural versus Urban Power on a Great Plains Frontier* (Westport, CT: Greenwood Press, 1973), 29; and Seth K. Humphrey, *Following the Prairie Frontier* (Minneapolis: University of Minnesota Press, 1931), 171, 116.

8. Allan G. Bogue, "Land Credit for Northern Farmers, 1789–1940," *Agricultural History* 50 (January 1976): 81; Bogue, *Money at Interest: The Farm Mortgage on the Middle Border* (Lincoln: University of Nebraska Press, 1955), 99; I. E. M. Smith, "Two City Girls' Experience in Holding Down a Claim: A Montana Pastoral," *Overland Monthly* 24 (August 1894): 160.

9. Bogue, *Money at Interest*, 214; Shannon, *The Farmer's Last Frontier*, 42; Richard C. Overton, *Burlington West: A Colonization History of the Burlington Railroad* (Cambridge: Harvard University Press, 1941), 314–315, 446–447; James B. Hedges, "The Colonizing Work of the Northern Pacific Railroad," *Mississippi Valley Historical Review* 13 (1926): 321.

10. Hinman and Rankin, *Eleven Southeastern Nebraska Townships*, 39, 49; Bogue, "Land Credit for Northern Farmers," 81; Scott G. and Sally McNall, *Plains Families: Exploring Sociology through Social History* (New York: St.Martin's Press, 1983), 50.

11. Hinman and Rankin, *Eleven Southeastern Nebraska Townships*, 38.

12. Bogue, *Money at Interest*, 7, 124; Hinman and Rankin, *Eleven Southeastern Nebraska Townships*, 39.

13. Bogue, *Money at Interest*, 129; Kenneth A. Snowden, "Covered Farm Mortgage Bonds in the United States during the Late Nineteenth Century," *Journal of Economic History* 70 (December 2010): 788.

14. Bogue, *Money at Interest*, 85; Hinman and Rankin, *Eleven Southeastern Nebraska Townships*, 41; Snowden, "Covered Farm Mortgage Bonds," 784–785, 793, 807.

15. Douglas C. North, Terry L. Anderson, and Peter J. Hill, *Growth and Welfare in the American Past: A New Economic History* (Englewood Cliffs, NJ: Prentice-Hall, 1983), 130; Bogue, *Money at Interest*, 16, 23; Hinman and Rankin, *Eleven Southeastern Nebraska Townships*, 32–33.

16. Bogue, *Money at Interest*, 16, 84.

17. Hinman and Rankin, *Eleven Southeastern Nebraska Townships*, 17, 32; Svendsbye, *I Paid All My Debts*, 40.

18. Bogue, *Money at Interest*, 114; Hinman and Rankin, *Eleven Southeastern Nebraska Townships*, 24.

19. Bogue, *Money at Interest*, 275; Mari Sandoz, *Old Jules* (Boston: Little, Brown, 1935), 308.

20. Bogue, *Money at Interest*, 115; Hinman and Rankin, *Eleven Southeastern Nebraska Townships*, 18.

21. Jon K. Lauck, *Prairie Republic: The Political Culture of Dakota Territory, 1879–1889* (Norman: University of Oklahoma Press, 2010), 167; Craig Miner, "A Place of Boom and Bust: Hard Times Come to Kansas," *Kansas History* 34 (Spring 2011): 76.

22. Bogue, *Money at Interest*, 121; Rezin W. McAdam, "The Peopling of the Plains," *Overland Monthly* 42 (August 1903): 134; Charles Moreau Harger, "Short Grass Country," *Harpers Weekly* 45 (January 26, 1901): 88.

23. Humphrey, *Following the Prairie Frontier*, 95.

24. Katherine Harris, *Long Vistas: Women and Families on Colorado Homesteads* (Niwot, CO: University Press of Colorado, 1993), 40; Craig Miner, *Next Year Country: Dust to Dust in Western Kansas, 1890–1940* (Lawrence: University Press of Kansas, 2006), 11; Humphrey, *Following the Prairie Frontier*, 165.

25. A. E. Dickey, "Modern Pioneer," *World To-Day* 14 (February 1908): 201; Bogue, *Money at Interest*, 251; Paul W. Gates, *History of Public Land Law Development* (Washington, D.C.: Government Printing Office, 1969), 213.

26. Bogue, *Money at Interest*, 50, 90.

27. Charles Moreau Harger, "To-Day's Chance for the Western Settler," *Outlook* 78 (17 December 1904): 981; Hinman and Rankin, *Eleven Southeastern Nebraska Townships*, 37.

CHAPTER FOUR: How They Built Communities

1. Walter Prescott Webb, *The Great Plains* (Boston: Ginn, 1931), 8.

2. Paula M. Nelson, *After the West Was Won: Homesteaders and Town-Builders in Western South Dakota, 1900–1917* (Iowa City: University of Iowa Press, 1986), 21–22.

3. Jon K. Lauck, *Prairie Republic: The Political Culture of Dakota Territory, 1879–1889* (Norman: University of Oklahoma Press, 2010); Bruce Garver, "Immigration to the Great Plains, 1865–1914: War, Politics, Technology, and Economic Development," *Great Plains Quarterly* 31 (Summer 2011): 186.

4. E. V. Smalley, "The Isolation of Life on Prairie Farms," *Atlantic* 72 (September 1893): 379; Charles Moreau Harger, "Short Grass Country," *Harpers Weekly* 45 (26 January 1901): 88.

5. Smalley, "Isolation of Life," 379; Mrs. Raymond Millbrook, ed., "Mrs. Hattie E. Lee's Story of Her Life in Western Kansas," *Kansas Historical Quarterly* 22 (1956): 125.

6. For inland towns and their functions, see John C. Hudson, *Plains Country Towns* (Minneapolis: University of Minnesota Press, 1985), 28, 123.

7. Lauck, *Prairie Republic*, 51.

8. Roderick Cameron, *Pioneer Days in Kansas: A Homesteader's Narrative of Early Settlement and Farm Development in the High Plains Country of Northwest Kansas* (Belleville, KS: Cameron Book, 1951), 44.

9. Nathan B. Sanderson, "More than a Potluck: Shared Meals and Community Building in Rural Nebraska at the Turn of the Twentieth Century," *Nebraska History* 89 (Fall 2008): 120–131. For the role of women in creating and sustaining social net-

works on the Great Plains see esp. Barbara Handy-Marchello, *Women of the Northern Plains: Gender and Settlement on Homestead Frontier, 1870–1930* (St. Paul: Minnesota Historical Society Press, 2005), 85–115.

10. J. Sanford Rikoon, *Rachel Calof's Story: Jewish Homesteader on the Northern Plains* (Bloomington: Indiana University Press, 1995), 71, 85.

11. Sandra L. Myres, *Westering Women and the Frontier Experience, 1800–1915* (Albuquerque: University of New Mexico Press, 1982), 186.

12. Faye C. Lewis, *Nothing to Make a Shadow* (Ames: Iowa State University Press, 1971), 82; Robert C. Ostergren, *A Community Transplanted: The Trans-Atlantic Experience of a Swedish Immigrant Settlement in the Upper Middle West, 1835–1915* (Madison: University of Wisconsin Press, 1988), 211.

13. Mary W. M. Hargreaves, "Rural Education on the Northern Plains Frontier," *Journal of the West* 18 (October 1979): 30.

14. Paul W. Gates, *History of Public Land Law Development* (Washington: U.S. Government Printing Office, 1968), 407; Allen G. Bogue, *Money at Interest: The Farm Mortgage on the Middle Border* (Lincoln: University of Nebraska Press, 1955), 168–169.

15. Marilyn Irwin Holt, *Children of Western Plains: The Nineteenth-Century Experience* (Chicago: Ivan R. Dee, 2003), 93; Myres, *Westering Women*, 184; Nina Farley Wishek, *Along the Trails of Yesterday: A Story of McIntosh County* (Ashley, ND: Ashley Tribune, 1941), 180–181.

16. Cameron, *Pioneer Days in Kansas*, 49, 59.

17. For a survey of one-room school conditions and educational programs, see Wayne E. Fuller, *The Old Country School: The Story of Rural Education in the Middle West* (Chicago: University of Chicago Press, 1982).

18. Cameron, *Pioneer Days in Kansas*, 84–85; Mary Hurlbut Cordier, *Schoolwomen of the Prairies and Plains: Personal Narratives from Iowa, Kansas, and Nebraska, 1860s—1920s* (Albuquerque: University of New Mexico Press, 1992), 90.

19. For the range of schoolhouse-centered activities in one community, see Lewis, *Nothing to Make a Shadow*, 73–74.

20. Scott G. and Sally Allen McNall, *Plains Families: Exploring Sociology through Social History* (New York: St. Martin's Press, 1983), 49.

21. My understanding of the relationship between the railroads and infrastructure development on the plains draws heavily on Hudson's *Plains Country Towns*.

22. James C. Malin, "The Adaptation of the Agricultural System to Subhumid Environment," in *History and Ecology: Studies of the Grassland*, ed. Robert P. Swierenga (Lincoln: University of Nebraska Press, 1984), 161–162; Lloyd A. Svendsbye, *I Paid All My Debts . . . : A Norwegian-American Immigrant Saga of Life on the Prairie of North Dakota* (Minneapolis: Lutheran University Press, 2009), 50.

23. Carlos A. Schwantes and James P. Ronda, *The West the Railroads Made* (Seattle: University of Washington Press, 2008), 104; Dorothy Hubbard Schwieder, *Growing Up with the Town: Family and Community on the Great Plains* (Iowa City: University

of Iowa Press, 2002), 13; Hiram M. Drache, *The Challenge of the Prairie: Life and Times of Red River Pioneers* (Fargo: North Dakota Institute for Regional Studies, 1970), 149; Laura Ingalls Wilder, *The Long Winter* (New York: Harper Collins, 1953), 142–143.

24. Cameron, *Pioneer Days in Kansas*, 109.

25. Nelson, *After the West Was Won*, 85.

26. Hudson, *Plains Country Towns*, 12, 79, 100–101.

27. Schwieder, *Growing Up with the Town*, 33, 38; Lela Barnes, ed., "North Central Kansas in 1887–1889: From the Letters of Leslie and Susan Snow of Junction City," *Kansas Historical Quarterly* 39 (Autumn 1963): 281; Cameron, *Pioneer Days in Kansas*, 6; Seth K. Humphrey, *Following the Prairie Frontier* (Minneapolis: University of Minnesota Press, 1931), 78.

28. Richard C. Overton, *Burlington West: A Colonization History of the Burlington Railroad* (Cambridge: Harvard University Press, 1941), 182; Hudson, *Plains Country Towns*, 12, 100–101.

29. Hudson, *Plains Country Towns*, 95.

30. R. Alton Lee, *T-Town on the Plains* (Manhattan, KS: Sunflower University Press, 1999), 34–39. A fine local study of the influence of railroad decisions on local communities is Carroll Engelhardt's *Gateway to the Northern Plains: Railroads and the Birth of Fargo-Moorhead* (Minneapolis: University of Minnesota Press, 2007).

31. Jane Taylor Nelsen, ed., *A Prairie Populist: The Memoirs of Luna Kellie* (Iowa City: University of Iowa Press, 1992), 122; James C. Malin, "Rural Life and Subhumid Environment," in *History and Ecology: Studies of the Grassland*, ed. Robert P. Swierenga (Lincoln: University of Nebraska Press, 1984), 203; Schwantes and Ronda, *The West the Railroads Made*, 107. For a recent treatment of the economic weaknesses of the railroad on the plains, see Richard White, *Railroaded: The Transcontinentals and the Making of Modern America* (New York: W.W. Norton, 2011).

32. Hudson, *Plains Country Towns*, 62–64.

33. My understanding of the lumber business on the Great Plains is drawn largely from John N. Vogel's *Great Lakes Lumber on the Great Plains: The Laird, Norton Lumber Company in South Dakota* (Iowa City: University of Iowa Press, 1992).

34. Malin, "Rural Life and Subhumid Environment," 200.

35. Glenda Riley, *The Female Frontier: A Comparative View of Women on the Prairie and the Plains* (Lawrence: University Press of Kansas, 1988), 91; David Debert Kruger, "Main Street Empire: J C Penney in Nebraska," *Nebraska History* 92 (Summer 2011): 540–569.

36. Hudson, *Plains Country Towns*, 106–107; Handy-Marchello, *Women of the Northern Plains*, 127.

37. Schwieder, *Growing Up with the Town*, 47.

38. Riley, *The Female Frontier*, 175–181.

39. Venola Lewis Bivans, ed., "The Diary of Luna E. Warner, a Kansas Teenager of the Early 1870s," *Kansas Historical Quarterly* 35 (1969): 286–287.

40. Ibid.

41. Nelson, *After the West Was Won*, 102; Hudson, *Plains Country Towns*, 125; Stanley B. Parsons, *The Populist Context: Rural versus Urban Power on a Great Plains Frontier* (Westport, CT: Greenwood Press, 1973), 46.

CHAPTER FIVE: How the Plains Matured

1. Sarah Comstock, "The Kansas Farmer's Wife," *Collier's* 42 (2 January 1909): 8.
2. Beatrice Roth Gant, ed., "Homesteading in Meade County, 1910–1911: The Memoir of William L. Roth," *South Dakota History* 40 (Fall 2010): 243–255.
3. Walter Nugent, *Into the West: The Story of Its People* (New York: Alfred A. Knopf, 1999), 131.
4. For the Golden Age of Agriculture, see David B. Danbom, *Born in the Country: A History of Rural America*, 2nd ed. (Baltimore: Johns Hopkins University Press, 2006), 163–168.
5. Charles Moreau Harger, "To-Day's Chance for the Western Settler," *Outlook* 78 (17 December 1904): 980; R. Douglas Hurt, *The Big Empty: The Great Plains in the Twentieth Century* (Tucson: University of Arizona Press, 2011), 2.
6. Katherine Harris, *Long Vistas: Women and Families on Colorado Homesteads* (Niwot, CO: University Press of Colorado, 1993), 65, 142; Nugent, *Into the West*, 144.
7. Paula Nelson, *After the West Was Won: Homesteaders and Town-Builders in Western South Dakota, 1900–1917* (Iowa City: University of Iowa Press, 1986), 30; H. Elaine Lindgren, *Land in Her Own Name* (Fargo: North Dakota Institute for Regional Studies), 36; Charles Moreau Harger, "The Land Movement and Western Finance," *North American Review* 192 (December 1910): 746.
8. Harris, *Long Vistas*, 58–59, 77.
9. Mary Boynton Cowdrey, ed., *The Checkered Years*, by Mary Dodge Woodward (Cass County Historical Society, n.d.), 242; James I. Stewart, "Economic Opportunity or Hardship? The Causes of Geographic Mobility on the Agricultural Frontier, 1860–1880," *Journal of Economic History* 69 (March 2009): 264.
10. Gilbert C. Fite, *The Farmer's Frontier, 1865–1900* (New York: Holt, Rinehart and Winston, 1966): 45.
11. Mari Sandoz, *Old Jules* (Boston: Little, Brown, 1935), 180.
12. Charles Moreau Harger, "Short Grass Country," *Harper's Weekly* 45 (26 January 1901): 89; June Granatir Alexander, *Daily Life in Immigrant America, 1870–1920: How the Second Great Wave of Immigrants Made Their Way in America* (Chicago: Ivan R. Dee, 2009), 58; Carol K. Coburn, *Life at Four Corners: Religion, Gender, and Education in a German-Lutheran Community, 1868–1945* (Lawrence: University Press of Kansas, 1992), 16, 18; Stewart, "Economic Opportunity or Hardship?," 267–268.
13. Deborah Fink, *Agrarian Women: Wives and Mothers in Rural Nebraska, 1880–1940* (Chapel Hill: University of North Carolina Press, 1992), 53; Charles Postel, *The Populist Vision* (New York: Oxford University Press, 2007), 77; Carl F. Kraenzel, Watson Thomson, and Glenn H. Craig, *The Northern Plains in a World of Change* (Toronto?: Gregory-Cartwright, 1942), 50, 71.

14. For immigrant attitudes regarding American cultural, social, and economic values, see Jon Gjerde, *Minds of the West: Ethnocultural Evolution in the Rural Middle West, 1830–1917* (Chapel Hill: University of North Carolina Press, 1999).

15. Harger, "Land Movement and Western Finance," 753. For the Campbell Dry Farming System, see Mary W. M. Hargreaves, *Dry Farming in the Northern Plains, 1900–1925* (Cambridge: Harvard University Press, 1957).

16. For the railroads' efforts see Roy V. Scott, *Railroad Development Programs in the Twentieth Century* (Ames: Iowa State University Press, 1985), and Claire Strom, *Profiting from the Plains: The Great Northern Railway and Corporate Development of the American West* (Seattle: University of Washington Press, 2003).

17. Heather Cox Richardson, *The Greatest Nation of the Earth: Republican Economic Policies during the Civil War* (Cambridge: Harvard University Press, 1997), 139–178. For the activities and accomplishments of one agricultural experiment station on the Great Plains, see David B. Danbom, *"Our Purpose Is to Serve": The First Century of the North Dakota Agricultural Experiment Station* (Fargo: North Dakota Institute for Regional Studies, 1990).

18. Harger, "Today's Chance for the Western Settler," 982.

19. Robert Wuthnow, *Remaking the Heartland: Middle America since the 1950s* (Princeton: Princeton University Press, 2011), 29.

20. Fite, *The Farmer's Frontier*, 53.

21. Alan L. Olmstead and Paul W. Rhode, "The Red Queen and the Hard Reds: Productivity Growth in American Wheat, 1800–1940," *Journal of Economic History* 62 (December 2002): 929–966.

22. A. E. Dickey, "Modern Pioneer," *World To-Day* 14 (February 1908): 204–205. For housing see Fred W. Peterson, *Homes in the Heartland: Balloon Frame Farmhouses of the Upper Midwest, 1850–1920* (Lawrence: University Press of Kansas, 1992), 196.

23. For the Country Life Movement see David B. Danbom, *The Resisted Revolution: Urban America and the Industrialization of Agriculture, 1900–1930* (Ames: Iowa State University Press, 1979).

24. Eleanor H. Hinman and J. O. Rankin, *Farm Mortgage History of Eleven Southeastern Nebraska Townships, 1870–1932* (Lincoln: College of Agriculture, University of Nebraska, Agricultural Experiment Station Research Bulletin 67, August 1933), 10; Christopher S. Decker and Daniel T. Flynn, "The Railroad's Impact on Land Values in the Upper Great Plains at the Closing of the Frontier," *Historical Methods* 40 (Winter 2007): 33. For retirement strategies of farmers, see Chulee Lee, "Farm Values and Retirement of Farm Owners in Early-Twentieth-Century America," *Explorations in Economic History* 36 (1999): 387–408.

SELECTED FURTHER READING

The classic work on the Great Plains is Walter Prescott Webb's *The Great Plains* (Boston, 1931). Other older studies include Carl Frederick Kraenzel, Watson Thomson, and Glenn H. Craig, *The Northern Plains in a World of Change* (Toronto?, 1942); Fred A. Shannon, *The Farmer's Last Frontier: Agriculture, 1860–1879* (New York, 1945); Gilbert C. Fite, *The Farmer's Frontier, 1865–1900* (New York, 1966); and Everett Dick, *The Sod-House Frontier: A Social History of the Northern Plains from the Creation of Kansas and Nebraska to the Admission of the Dakotas* (Lincoln, NE, 1979). More recent treatments include Ian Frazier, *The Great Plains* (New York, 1989); Walter Nugent, *Into the West: The Story of Its People* (New York, 1999); Geoff Cunfer, *On the Great Plains: Agriculture and Environment* (College Station, TX, 2005); and R. Douglas Hurt, *The Big Empty: The Great Plains in the Twentieth Century* (Tucson, 2011). Two relatively recent books on government policies that helped stimulate European and European American settlement on the plains are Heather Cox Richardson, *The Greatest Nation of the Earth: Republican Economic Policies during the Civil War* (Cambridge, MA, 1997), and Stephen J. Rockwell, *Indian Affairs and the Administrative State in the Nineteenth Century* (Cambridge, U.K., 2010).

Historians of previous generations produced essential studies of federal land policy, including Benjamin Horace Hibbard, *A History of Public Land Policies* (New York, 1924), and Roy M. Robbins, *Our Landed Heritage: The Public Domain* (Princeton, NJ, 1942). The preeminent expert on land policy was Paul Wallace Gates, whose *Fifty Million Acres: Conflicts Over Kansas Land Policy, 1854–1890* (New York, 1966), and *History of Public Land Development* (Washington, D.C., 1968) remain essential. More recent, and limited, treatments include Hildegard Binder Johnson, *Order upon the Land: The U.S. Rectangular Land Survey and the Upper Mississippi Country* (New York, 1976), and John Opie, *The Law of the Land: Two Hundred Years of American Farmland Policy* (Lincoln, NE, 1987).

Numerous studies touch on the significance of the railroad to the settlement of the plains. Valuable recent works include Claire Strom, *Profiting from the Plains: The Great Northern Railway and the Corporate Development of the American West* (Seattle, 2003); Carlos A. Schwantes and James P. Ronda, *The West the Railroads Made* (Seattle, 2008); and Richard White, *Railroaded: The Transcontinentals and the Making of Modern America* (New York, 2011). A good study of how one railroad populated its service area is Richard C. Overton, *Burlington West: A Colonization History of the Burlington Railroad* (Cambridge, MA, 1941). John C. Hudson's *Plains Country Towns* (Minneapolis, 1985) explores the town-development activities of the railroads, and Roy V. Scott, *Railroad Development Programs in the Twentieth Century* (Ames, IA, 1985), details their efforts to promote agriculture.

Various aspects of the farm-making process are addressed by the Great Plains historians mentioned above, as well as by Mary Wilma M. Hargreaves, *Dry Farming in the Northern Great Plains, 1900–1925* (Cambridge, MA, 1957); Hiram M. Drache, *The Challenge of the Prairie: Life and Times of Red River Pioneers* (Fargo, ND, 1970); and Fred W. Peterson, *Homes in the Heartland: Balloon Frame Farmhouses of the Upper Midwest, 1850–1920* (Lawrence, KS, 1992). Popular authors such as Willa Cather, Hamlin Garland, and Laura Ingalls Wilder also address farm-making issues in some of their work. Settlers' reminiscences present valuable portrayals of the travails and triumphs of the effort to make farms and homes on the Great Plains. Most valuable to me were J. Sanford Rikoon, ed., *Rachel Calof's Story: Jewish Homesteader on the Northern Plains* (Bloomington, IN, 1995); Roderick Cameron, *Pioneer Days in Kansas: A Homesteader's Narrative of Early Settlement and Farm Development on the High Plains Country of Northwest Kansas* (Belleville, KS, 1951); Jane Taylor Nelsen, ed., *A Prairie Populist: The Memoirs of Luna Kellie* (Iowa City, 1992); Mari Sandoz, *Old Jules* (Boston, 1935); Lloyd A. Svendsbye, *I Paid All My Debts . . . : A Norwegian American Immigrant Saga of Life on the Prairie of North Dakota* (Minneapolis, 2009); and Nina Farley Wishek, *Along the Trails of Yesterday: A Story of McIntosh County* (Ashley, ND, 1941). Other memoirs and collections I found useful included L. L. Bloomenshine, *Prairie around Me* (San Diego, 1972); Mary Hurlbut Cordier, *Schoolwomen of the Prairies and Plains: Personal Narratives from Iowa, Kansas, and Nebraska, 1860s–1920s* (Albuquerque, 1992); Baldwin F. Kruse, *Paradise on the Prairie: Nebraska Settler Stories* (Lincoln, NE, 1986); Faye C. Lewis, *Nothing to Make a Shadow* (Ames, IA, 1971); Scott G. and Sally Allen McNall, *Plains Families: Exploring Sociology through Social History* (New York, 1983); Mary Dodge Woodward, *The Checkered Years* (Fargo, ND, no date); and Walker D. Wyman, *Frontier Woman: The Life of a Woman Homesteader on the Dakota Frontier* (River Falls, WI, 1972).

The farm credit system on the plains has received relatively little attention from historians. The best study of the subject remains Allan G. Bogue's *Money at Interest: The Farm Mortgage on the Middle Border* (Lincoln, NE, 1955). An old but valuable community study is Eleanor H. Hinman and J. O. Rankin, *Farm Mortgage History of Eleven Southeastern Nebraska Townships, 1870–1932* (Lincoln, NE, 1933). For a more personal view, see the memoir of mortgage company representative Seth K. Humphrey, *Following the Prairie Frontier* (Minneapolis, 1931).

For town building see Hudson's *Plains Country Towns* and Lewis Atherton's classic, *Main Street on the Middle Border* (Bloomington, IN, 1954). Two valuable town studies are R. Alton Lee, *T-Town on the Plains* (Manhattan, KS, 1999), and Dorothy Hubbard Schwieder, *Growing Up with the Town: Family and Community on the Great Plains* (Iowa City, 2002). For common schools see Wayne E. Fuller, *The Old Country School: The Story of Rural Education in the Middle West* (Chicago, 1982), and Paul Theobald, *Call School: Rural Education in the Midwest to 1918* (Carbondale, IL, 1995). A recent investigation of the development of political culture is Jon K. Lauck, *Prairie Republic: The Political Culture of Dakota Territory, 1879–1889* (Norman, OK, 2010).

Some insightful studies of the immigrant experience on the plains include Robert C. Ostergren, *A Community Transplanted: The Trans-Atlantic Experience of a Swedish Immigrant Settlement in the Upper Middle West, 1835–1915* (Madison, WI, 1988); Carol K. Coburn, *Life at Four Corners: Religion, Gender, and Education in a German-Lutheran Community, 1868–1945* (Lawrence, KS, 1992); and June Granatir Alexander, *Daily Life in Immigrant America: How the Second Great Wave of Immigrants Made Their Way in America* (Chicago, 2009).

The historical literature on women on the plains, both as wives and as single settlers, is extensive. Among the most significant works are Joanna L. Stratton, *Pioneer Women: Voices from the Kansas Frontier* (New York, 1981); Sandra L. Myres, *Westering Women and the Frontier Experience, 1800–1915* (Albuquerque, 1982); Glenda Riley, *The Female Frontier: A Comparative View of Women on the Prairie and the Plains* (Lawrence, KS, 1988); H. Elaine Lindgren, *Land in Her Own Name: Women as Homesteaders in North Dakota* (Fargo, ND, 1991); Deborah Fink, *Agrarian Women: Wives and Mothers in Rural Nebraska, 1880–1940* (Chapel Hill, NC, 1992); Katherine Harris, *Long Vistas: Women and Families on Colorado Homesteads* (Niwot, CO, 1993); Linda Peavy and Ursula Smith, *Pioneer Women: The Lives of Women on the Frontier* (Norman, OK, 1996); and Barbara Handy-Marchello, *Women of the Northern Plains: Gender and Settlement on the Homestead Frontier* (St. Paul, MN, 2005).

Many of the books I've mentioned deal at least in part with settlement of the western plains. More specific treatments include Paula M. Nelson, *After the West Was Won: Homesteaders and Town-Builders in Western South Dakota, 1900–1917* (Iowa City, 1986), and Craig Miner, *Next Year Country: Dust to Dust in Western Kansas, 1890–1940* (Lawrence, KS, 2006).

INDEX